Textbook of Soil Science

Textbook of Soil Science

Edited by
Alfie Williams

Larsen & Keller
www.larsen-keller.com

Textbook of Soil Science
Edited by Alfie Williams
ISBN 978-1-63549-001-5 (Hardback)

Larsen & Keller

Published by Larsen and Keller Education,
5 Penn Plaza,
19th Floor,
New York, NY 10001, USA

Cataloging-in-Publication Data

Textbook of soil science / edited by Alfie Williams.
p. cm.
Includes bibliographical references and index.
ISBN 978-1-63549-001-5
1. Soil science--Textbooks. 2. Soils--Study and teaching--Textbooks.
I. Williams, Alfie.
S591.5 .T49 2017
631.4--dc23

The publisher's policy is to use permanent paper from mills that operate a sustainable forestry policy. Furthermore, the publisher ensures that the text paper and cover boards used have met acceptable environmental accreditation standards.

Printed and bound in China.

For more information regarding Larsen and Keller Education and its products, please visit the publisher's website www.larsen-keller.com

Table of Contents

Preface

Soil Science is a primary aspect of agriculture. This branch of science is not only helpful in determining the suitable crops for farming but also works as a beneficiary agent in enhancing the quality of the soil and the yield. This textbook gives an overview of significant areas of this field such as soil formation, classification and mapping along with the physical, chemical and biological properties of the soil. This book will appeal to readers from several scientific disciplines and serve as a good source of knowledge.

A short introduction to every chapter is written below to provide an overview of the content of the book:

Chapter 1-Soil science is a primary area of study under agricultural science. It plays a crucial role in the production of crops. This elaborate chapter provides an all-encompassing introduction to the vast field of soil science; **Chapter 2**-Soil is regarded as one of the most crucial factor affecting crop yield. This chapter aims to provide a comprehensive overview of soil by detailing its various layers such as topsoil, etc. it also glances upon the topic of soil structure to further facilitate the understanding of the subject; **Chapter 3**-This chapter will provide a comprehensive overview of all the major types of soil which includes alkali soil, clay and serpentine soil. It also glances upon the process of classification of soil and the factors influencing it. The chapter strives to provide a clear understanding about the types of soil in a systematic yet critical manner; **Chapter 4**-Soil depicts various characteristics and properties which are significant for a complete understanding of soil science. Soil type, Soil pH, Soil fertility, etc. are some of the important topics related to soil science which are rigorously examined in this chapter. It unfolds these significant aspects in a crucial manner; **Chapter 5**-The process of soil formation is scientifically known as pedogenesis. This chapter lays special emphasis on this process by explaining the minutest details of this process. It also highlights the factors affecting this process. This chapter will help readers broaden their perspective in the field of soil science; **Chapter 6**-This chapter primarily focuses on soil morphology and soil horizon which are important facets of soil science. The chapter aims to shed light on the unexplored aspects of soil morphology to provide a better understanding of the subject matter; **Chapter 7**-Soil survey is one of the most significant aspect related to soil science. This chapter will unfold this topic in an easy and comprehensible manner. It will also cover some related topics like pedotransfer functions, catena (soil), etc. to enhance the knowledge of the topic; **Chapter 8** - Soil science can primarily be divided into two main branches–edaphology and pedology. These branches explain the relationship of soil with its environment. Both the branches have been thoroughly discussed in this chapter to provide an in-depth understanding of soil science; **Chapter 9**-This chapter provides a comprehensive insight into the mechanisms of soil variables and soil mechanics which contribute to the broadening of the existing knowledge of soil science. This chapter also incorporates related aspects like soil acidification, soil gas, soil texture, etc. and elaborates their role in soil science; **Chapter 10**-One of the primary fields of soil science is soil ecology. This chapter studies in detail the processes and procedures of soil ecology and soil biodiversity providing a detailed account on each of these aspects related to the specified area of study; **Chapter 11**-Soil erosion and its prevention is a major area of study under soil science. The following chapter focuses on the varied reasons for soil contamination along with describing soil conservation processes as well. It also glances at land degradation with respect to soil. It will help the readers develop a deeper understanding of the phenomena of soil loss and contamination; **Chapter 12**-The following chapter will emphasize on the different scientific aspects of soil. The chapter has elaborated and discussed at length, multiple integrated areas of study like soil physics, soil chemistry and soil biology which are closely associated with the umbrella field of soil science; **Chapter 13**-The tools and techniques employed to study soil are of great importance. The following chapter will provide a comprehensive insight into the major tools and techniques that are used in the subject of soil science such as soil nailing, pedometrics, soil compaction, etc. They have been discussed elaborately so as to give the readers a complete understanding of soil science.

I extend my sincere thanks to the publisher for considering me worthy of this task. Finally, I thank my family for being a source of support and help.

Editor

1

An Introduction to Soil Science

Soil science is a primary area of study under agricultural science. It plays a crucial role in the production of crops. This elaborate chapter provides an all-encompassing introduction to the vast field of soil science.

Soil Science

A sylviculturist, at work

Soil science is the study of soil as a natural resource on the surface of the Earth including soil formation, classification and mapping; physical, chemical, biological, and fertility properties of soils; and these properties in relation to the use and management of soils.

Sometimes terms which refer to branches of soil science, such as pedology (formation, chemistry, morphology and classification of soil) and edaphology (influence of soil on organisms, especially plants), are used as if synonymous with soil science. The diversity of names associated with this discipline is related to the various associations concerned. Indeed, engineers, agronomists, chemists, geologists, physical geographers, ecologists, biologists, microbiologists, sylviculturists, sanitarians, archaeologists, and specialists in regional planning, all contribute to further knowledge of soils and the advancement of the soil sciences.

Soil scientists have raised concerns about how to preserve soil and arable land in a world with a growing population, possible future water crisis, increasing per capita food consumption, and land degradation.

Fields of Study

Soil occupies the pedosphere, one of Earth's spheres that the geosciences use to organize the Earth conceptually. This is the conceptual perspective of pedology and edaphology, the two main branches of soil science. Pedology is the study of soil in its natural setting. Edaphology is the study of soil in relation to soil-dependent uses. Both branches apply a combination of soil physics, soil chemistry, and soil biology. Due to the numerous interactions between the biosphere, atmosphere and hydrosphere that are hosted within the pedosphere, more integrated, less soil-centric concepts are also valuable. Many concepts essential to understanding soil come from individuals not identifiable strictly as soil scientists. This highlights the interdisciplinary nature of soil concepts.

Research

Dependence on and curiosity about soil, exploring the diversity and dynamics of this resource continues to yield fresh discoveries and insights. New avenues of soil research are compelled by a need to understand soil in the context of climate change, greenhouse gases, and carbon sequestration. Interest in maintaining the planet's biodiversity and in exploring past cultures has also stimulated renewed interest in achieving a more refined understanding of soil.

Mapping

Most empirical knowledge of soil in nature comes from soil survey efforts. Soil survey, or soil mapping, is the process of determining the soil types or other properties of the soil cover over a landscape, and mapping them for others to understand and use. It relies heavily on distinguishing the individual influences of the five classic soil forming factors. This effort draws upon geomorphology, physical geography, and analysis of vegetation and land-use patterns. Primary data for the soil survey are acquired by field sampling and supported by remote sensing.

Classification

As of 2006, the World Reference Base for Soil Resources, via its Land & Water Development division, is the pre-eminent soil classification system. It replaces the previous FAO soil classification.

The WRB borrows from modern soil classification concepts, including USDA soil taxonomy. The classification is based mainly on soil morphology as an expression pedogenesis. A major difference with USDA soil taxonomy is that soil climate is not part of the system, except insofar as climate influences soil profile characteristics.

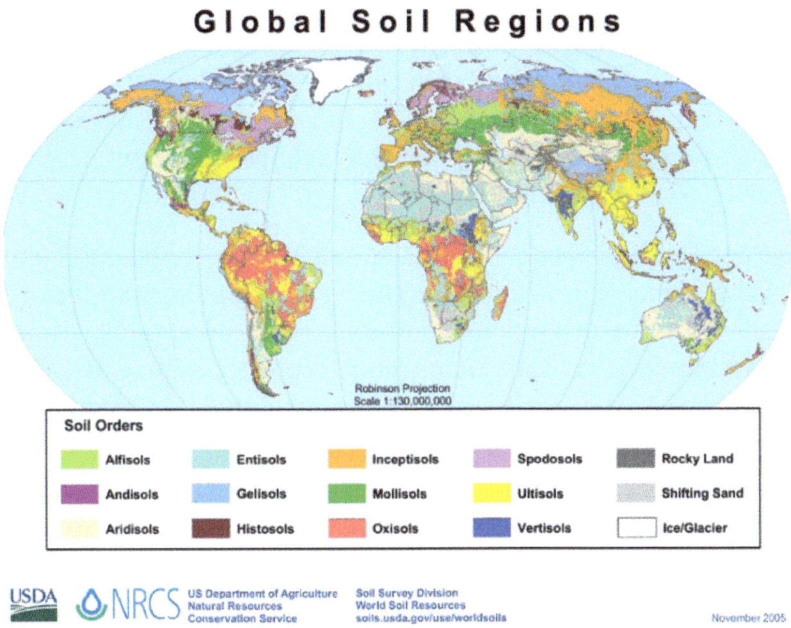

Map of global soil regions from the USDA

Many other classification schemes exist, including vernacular systems. The structure in vernacular systems are either nominal, giving unique names to soils or landscapes, or descriptive, naming soils by their characteristics such as red, hot, fat, or sandy. Soils are distinguished by obvious characteristics, such as physical appearance (e.g., color, texture, landscape position), performance (e.g., production capability, flooding), and accompanying vegetation. A vernacular distinction familiar to many is classifying texture as heavy or light. Light soil content and better structure, take less effort to turn and cultivate. Contrary to popular belief, light soils do not weigh less than heavy soils on an air dry basis nor do they have more porosity.

History

Vasily Dokuchaev, a Russian geologist, geographer and early soil scientist, is credited with identifying soil as a resource whose distinctness and complexity deserved to be separated conceptually from geology and crop production and treated as a whole.

Previously, soil had been considered a product of chemical transformations of rocks, a dead substrate from which plants derive nutritious elements. Soil and bedrock were in fact equated. Dokuchaev considers the soil as a natural body having its own genesis and its own history of development, a body with complex and multiform processes taking place within it. The soil is considered as different from bedrock. The latter becomes soil under the influence of a series of soil-formation factors (climate, vegetation, country, relief and age). According to him, soil should be called the "daily" or outward horizons of rocks regardless of the type; they are changed naturally by the common effect of water, air and various kinds of living and dead organisms.

A 1914 encyclopedic definition: "the different forms of earth on the surface of the rocks, formed by the breaking down or weathering of rocks". serves to illustrate the historic view of soil which persisted from the 19th century. Dokuchaev's late 19th century soil concept developed in the 20th century to one of soil as earthy material that has been altered by living processes. A corollary concept is that soil without a living component is simply a part of earth's outer layer.

Further refinement of the soil concept is occurring in view of an appreciation of energy transport and transformation within soil. The term is popularly applied to the material on the surface of the Earth's moon and Mars, a usage acceptable within a portion of the scientific community. Accurate to this modern understanding of soil is Nikiforoff's 1959 definition of soil as the "excited skin of the sub aerial part of the earth's crust".

Areas of Practice

Academically, soil scientists tend to be drawn to one of five areas of specialization: microbiology, pedology, edaphology, physics or chemistry. Yet the work specifics are very much dictated by the challenges facing our civilization's desire to sustain the land that supports it, and the distinctions between the sub-disciplines of soil science often blur in the process. Soil science professionals commonly stay current in soil chemistry, soil physics, soil microbiology, pedology, and applied soil science in related disciplines

One interesting effort drawing in soil scientists in the USA as of 2004 is the Soil Quality Initiative. Central to the Soil Quality Initiative is developing indices of soil health and then monitoring them in a way that gives us long term (decade-to-decade) feedback on our performance as stewards of the planet. The effort includes understanding the functions of soil microbiotic crusts and exploring the potential to sequester atmospheric carbon in soil organic matter. The concept of soil quality, however, has not been without its share of controversy and criticism, including critiques by Nobel Laureate Norman Borlaug and World Food Prize Winner Pedro Sanchez.

A more traditional role for soil scientists has been to map soils. Most every area in the United States now has a published soil survey, which includes interpretive tables as to how soil properties support or limit activities and uses. An internationally accepted soil taxonomy allows uniform communication of soil characteristics and functions. National and international soil survey efforts have given the profession unique insights into landscape scale functions. The landscape functions that soil scientists are called upon to address in the field seem to fall roughly into six areas:

- **Land-Based Treatment of Wastes**
    - Septic system
    - Manure
    - Municipal biosolids

    - Food and fiber processing waste
- **Identification and Protection of Environmentally Critical Areas**
    - Sensitive and unstable soils
    - Wetlands
    - Unique soil situations that support valuable habitat, and ecosystem diversity
- **Management for Optimum Land Productivity**
    - Silviculture
    - Agronomy
        - Nutrient management
        - Water management
    - Native vegetation
    - Grazing
- **Management for Optimum Water Quality**
    - Stormwater management
    - Sediment and erosion control
- **Remediation and Restoration of Damaged Lands**
    - Mine reclamation
    - Flood and storm damage
    - Contamination
- **Sustainability of Desired Uses**
    - Soil conservation

There are also practical applications of soil science that might not be apparent from looking at a published soil survey.

- Radiometric dating: specifically a knowledge of local pedology is used to date prior activity at the site
    - Stratification (archeology) where soil formation processes and preservative qualities can inform the study of archaeological sites
    - Geological phenomena
        - Landslides
        - Active faults
- **Altering Soils to Achieve New Uses**
    - Vitrification to contain radioactive wastes

    - Enhancing soil microbial capabilities in degrading contaminants (bioremediation).
    - Carbon sequestration
    - Environmental soil science
- Pedology
    - Soil genesis
    - Pedometrics
    - Soil morphology
        - Soil micromorphology
    - Soil classification
        - USDA soil taxonomy
- Soil biology
    - Soil microbiology
- Soil chemistry
    - Soil biochemistry
    - Soil mineralogy
- Soil physics
    - Pedotransfer function
    - Soil mechanics and engineering
- Soil hydrology, hydropedology

Fields of Application in Soil Science

- Climate change
- Ecosystem studies
- Pedotransfer function
- Soil fertility / Nutrient management
- Soil management
- Soil survey
- Standard methods of analysis
- Watershed and wetland studies

Related Disciplines

- Agricultural sciences
    - Agricultural soil science
    - Agrophysics science

    - Irrigation management
- Anthropology
    - archaeological stratigraphy
- Environmental science
    - Landscape ecology
- Physical geography
    - Geomorphology
- Geology
    - Biogeochemistry
    - Geomicrobiology
- Hydrology
    - Hydrogeology
- Waste management
- Wetland science

Soil Management

Soil management concerns all operations, practices, and treatments used to protect soil and enhance its performance.

Environment

According to the EPA, agricultural soil management practices can lead to production and emission of nitrous oxide (N_2O), a major greenhouse gas and air pollutant. Activities that can contribute to N_2O emissions include fertilizer usage, irrigation and tillage. The management of soils accounts for over half of the emissions from the Agriculture sector. Cattle livestocks account for one third of emissions, through methane emissions. Manure management and rice cultivation also emit emissions.

Methods that significantly enhance carbon sequestration in soil include no-till farming, residue mulching, cover cropping, and crop rotation, all of which are more widely used in organic farming than in conventional farming. Because only 5% of US farmland currently uses no-till and residue mulching, there is a large potential for carbon sequestration.

Practices

Soil management practices that affect soil quality:

- Controlling traffic on the soil surface helps to reduce soil compaction, which can reduce aeration and water infiltration.
- Cover crops keep the soil anchored and covered in off-seasons so that the soil is not eroded by wind and rain.
- Crop rotations for row crops alternate high-residue crops with lower-residue crops to increase the amount of plant material left on the surface of the soil during the year to protect the soil from erosion.
- Nutrient management can help to improve the fertility of the soil and the amount of organic matter content, which improves soil structure and function.
- Tillage, especially reduced-tillage or no-till operations limit the amount of soil disturbance while cultivating a new crop, and help to maintain plant residues on the surface of the soil for erosion protection and water retention.

Advantages of Soil Management

- Maintain soil fertility
- Restore soil fertility
- Make the agricultural process an economic one
- Help increase yield

References

- Jackson, J. A. (1997). Glossary of Geology (4. ed.). Alexandria, Virginia: American Geological Institute. p 604. ISBN 0-922152-34-9
- Buol, S. W.; Hole, F. D. & McCracken, R. J. (1973). Soil Genesis and Classification (First ed.). Ames, IA: Iowa State University Press. ISBN 978-0-8138-1460-5.
- Pielke, Roger (December 12, 2005). "Is Soil an Important Component of the Climate System?". The Climate Science Weblog. Archived from the original on 8 September 2006. Retrieved 19 April 2012.
- H. H. Janzen; et al. (2011). "Global Prospects Rooted in Soil Science". Soil Science Society of America Journal. 75 (1): 1. doi:10.2136/sssaj2009.0216.

A Comprehensive Study of Soil

2

Soil is regarded as one of the most crucial factor affecting crop yield. This chapter aims to provide a comprehensive overview of soil by detailing its various layers such as topsoil, etc. it also glances upon the topic of soil structure to further facilitate the understanding of the subject.

Soil

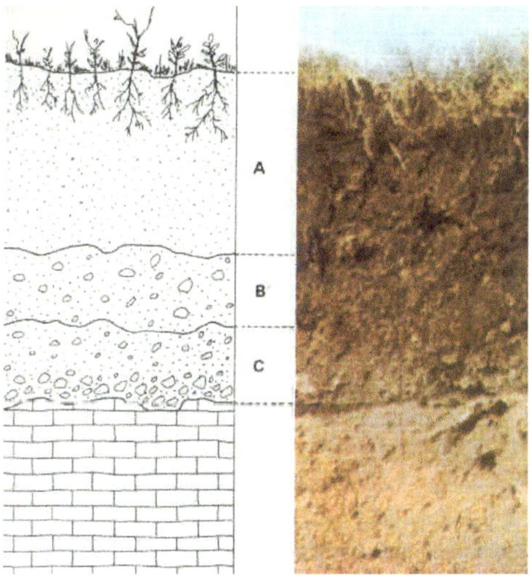

A represents soil; B represents laterite, a regolith; C represents saprolite, a less-weathered regolith; the bottom-most layer represents bedrock.

Soil is a mixture of minerals, organic matter, gases, liquids, and countless organisms that together support life on Earth. Soil is a natural body called the pedosphere which has four important functions: it is a medium for plant growth; it is a means of water storage, supply and purification; it is a modifier of Earth's atmosphere; it is a habitat for organisms; all of which, in turn, modify the soil.

Soil is called the "Skin of the Earth" and interfaces with its lithosphere, hydrosphere, atmosphere, and biosphere. The term pedolith, used commonly to refer to the soil, literally translates 'level stone'. Soil consists of a solid phase of minerals and organic matter, as well as a porous phase that holds gases and water. Accordingly, soils are often treated as a three-state system of solids, liquids, and gases.

Soil is a product of the influence of the climate, relief (elevation, orientation, and slope of terrain), organisms, and its parent materials (original minerals) interacting over time. Soil continually undergoes development by way of numerous physical, chemical and biological processes, which include weathering with associated erosion.

Most soils have a density between 1 and 2 g/cm³. Little of the soil of planet Earth is older than the Pleistocene and none is older than the Cenozoic, although fossilized soils are preserved from as far back as the Archean.

Surface-water-gley developed in glacial till, Northern Ireland

Soil science has two basic branches of study: edaphology and pedology. Edaphology is concerned with the influence of soils on living things. Pedology is focused on the formation, description (morphology), and classification of soils in their natural environment. In engineering terms, soil is referred to as regolith, or loose rock material that lies above the 'solid geology'. Soil is commonly referred to as "earth" or "dirt"; technically, the term "dirt" should be restricted to displaced soil.

As soil resources serve as a basis for food security, the international community advocates its sustainable and responsible use through different types of Soil Governance.

Overview

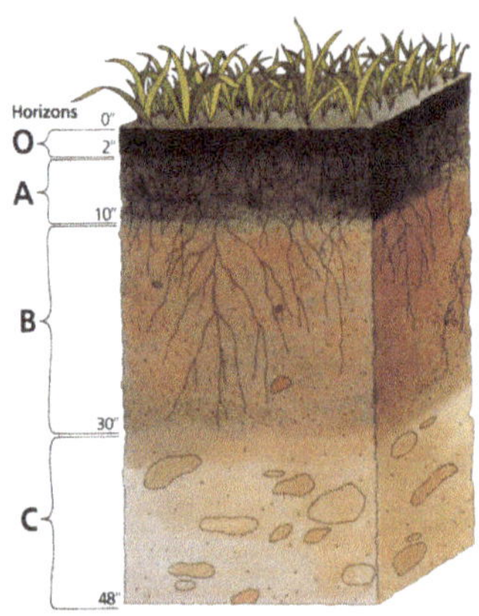

Soil Profile: Darkened topsoil and reddish subsoil layers are typical in some regions.

Soil is a major component of the Earth's ecosystem. The world's ecosystems are impacted in far-reaching ways by the processes carried out in the soil, from ozone depletion and global warming, to rainforest destruction and water pollution. Following the atmosphere, the soil is the next largest carbon reservoir on Earth, and it is potentially one of the most reactive to human disturbance and climate change. As the planet warms, soils will add carbon dioxide to the atmosphere due to its increased biological activity at higher temperatures. Thus, soil carbon losses likely have a large positive feedback response to global warming.

Soil acts as an engineering medium, a habitat for soil organisms, a recycling system for nutrients and organic wastes, a regulator of water quality, a modifier of atmospheric composition, and a medium for plant growth. Since soil has a tremendous range of available niches and habitats, it contains most of the Earth's genetic diversity. A gram of soil can contain billions of organisms, belonging to thousands of species. Soil has a mean prokaryotic density of roughly 10^{13} organisms per cubic meter, whereas the ocean has a mean prokaryotic density of roughly 10^{8} organisms per cubic meter. The carbon content of the soil is eventually returned to the atmosphere through the process of respiration carried out by heterotrophic organisms that feed upon the carbonaceous material in the soil. Since plant roots need oxygen, ventilation is an important characteristic of soil. This ventilation can be accomplished via networks of soil pores, which also absorb and hold rainwater making it readily available for plant uptake. Since plants require a nearly continuous supply of water, but most regions receive sporadic rainfall, the water-holding capacity of soils is vital for plant survival.

Soils can effectively remove impurities, kill disease agents, and degrade contaminants. Typically, soils maintain a net absorption of oxygen and methane, and undergo a net release of carbon dioxide and nitrous oxide. Soils offer plants physical support, air, water, temperature moderation, nutrients, and protection from toxins. Soils provide readily available nutrients to plants and animals by converting dead organic matter into various nutrient forms.

Soils supply plants with nutrients that are held in place by the clay and humus content of that soil. For optimum plant growth, the soil components by volume should be roughly 50% solids (45% mineral and 5% organic matter), and 50% voids of which half is occupied by water and half by gas. The percent soil mineral and organic content is typically treated as a constant, while the percent soil water and gas content is considered highly variable whereby a rise in one is simultaneously balanced by a reduction in the other. The pore space allows for the infiltration and movement of air and water, both of which are critical for life in soil. Compaction, a common problem with soils, reduces this space, preventing air and water from reaching the plant roots and soil organisms.

Given sufficient time, an undifferientated soil will evolve a soil profile which consists of two or more layers, referred to as soil horizons, that differ in one or more properties such as in their texture, structure, density, porosity, consistency, temperature, color, and reactivity. The horizons differ greatly in thickness and generally lack sharp boundaries. Soil profile development is dependent on the processes that form soils from their parent materials, the type of parent material, and the factors that control soil formation. The biological influences on soil properties are strongest near the surface, while the geochemical influences on soil properties increase with depth. Mature soil profiles in temperate climate regions typically include three basic master horizons: A, B and C. The solum normally includes the A and B horizons. The living component of the soil is largely confined to the solum. In the more hot, humid, climate of the tropics, a soil may have only a single horizon.

The soil texture is determined by the relative proportions of sand, silt, and clay in the soil. The addition of organic matter, water, gases and time causes the soil of that texture to develop into a larger soil structure called an aggregate. At that point a soil can be said to be developed, and can be described further in terms of color, porosity, consistency, reaction etc.

Of all the factors influencing the evolution of soil, water is the most powerful due to its involvement in the solution, erosion, transportation, and deposition of the materials of which a soil is composed. The mixture of water and the dissolved or suspended materials that occupy the soil pore space is called the soil solution. Since soil water is never pure water, but contains hundreds of dissolved organic and mineral substances, it may be more accurately called the soil solution. Water is central to the solution, precipitation and leaching of minerals from the soil profile. Finally, water affects the type of vegetation that grows in a soil, which in turn affects the development of the soil.

The most influential factor in stabilizing soil fertility are the soil colloids, clay and humus. Soil colloids behave as repositories of nutrients and moisture and so act to buffer the variations of soil solution ions and moisture. The contribution of soil colloids to soil nutrition are out of proportion to their part of the soil. Colloids act to store nutrients that might otherwise be leached from the soil or to release those ions in response to changes of soil pH, and so, make them available to plants.

The greatest influence on plant nutrient availability is soil pH, which is a measure of the hydrogen ion (acid-forming) soil reactivity, and is in turn a function of the soil materials, precipitation level, and plant root behavior. Soil pH strongly affects the availability of nutrients.

Most nutrients, with the exception of nitrogen, originate from minerals. Some nitrogen originates from rain (as dilute nitric acid), but most of the nitrogen is available in soils as a result of nitrogen fixation by bacteria. The action of microbes on organic matter and minerals may be to free nutrients for use, sequester them, or cause their loss from the soil by their volatilisation to gases or their leaching from the soil. The nutrients may be stored on soil colloids, or within live or dead organic matter, but they may not be accessible to plants due to extremes of pH.

The organic material of the soil has a powerful effect on its development, fertility, and available moisture. Following water and soil colloids, organic material is next in importance to a soil's formation and fertility.

History of The Study of Soil

Studies Concerning Soil Fertility

The history of the study of soil is intimately tied to our urgent need to provide food for ourselves and forage for our animals. Throughout history, civilizations have prospered or declined as a function of the availability and productivity of their soils.

The Greek historian Xenophon (450–355 B.C.) is credited with being the first to expound upon the merits of green-manuring crops: "But then whatever weeds are upon the ground, being turned into earth, enrich the soil as much as dung."

Columella's "Husbandry," circa 60 A.D., advocated the use of lime and that clover and alfalfa (green manure) should be turned under, and was used by 15 generations (450 years) under the Roman Empire until its collapse. From the fall of Rome to the French Revolution, knowledge of soil and agriculture was passed on from parent to child and as a result, crop yields were low. During the European Dark Ages, Yahya Ibn al-'Awwam's handbook, with its emphasis on irrigation, guided the people of North Africa, Spain and the Middle East; a translation of this work was finally carried to the southwest of the United States.

Experiments into what made plants grow first led to the idea that the ash left behind when plant matter was burned was the essential element but overlooked the role of nitrogen, which is not left on the ground after combustion. In about 1635, the Flemish chemist Jan Baptist van Helmont thought he had proved water to be the essential element from his famous five years' experiment with a willow tree grown with only the addition of rainwater. His conclusion came from the fact that the increase in the plant's weight had apparently been produced only by the addition of water, with no reduction in the soil's weight. John Woodward (d. 1728) experimented with various types of water ranging from clean to muddy and found muddy water the best, and so he concluded that earthy matter was the essential element. Others concluded it was humus in the soil that passed some essence to the growing plant. Still others held that the vital growth principal was something passed from dead plants or animals to the new plants. At the start of the 18th century, Jethro Tull demonstrated that it was beneficial to cultivate (stir) the soil, but his opinion that the stirring made the fine parts of soil available for plant absorption was erroneous.

As chemistry developed, it was applied to the investigation of soil fertility. The French chemist Antoine Lavoisier showed in about 1778 that plants and animals must "combust" oxygen internally to live and was able to deduce that most of the 165-pound weight of van Helmont's willow tree derived from air. It was the French agriculturalist Jean-Baptiste Boussingault who by means of experimentation obtained evidence showing that the main sources of carbon, hydrogen and oxygen for plants were the air and water. Justus von Liebig in his book *Organic Chemistry in its Applications to Agriculture and Physiology* (published 1840), asserted that the chemicals in plants must have come from the soil and air and that to maintain soil fertility, the used minerals must be replaced. Liebig nevertheless believed the nitrogen was supplied from the air. The enrichment of soil with guano by the Incas was rediscovered in 1802, by Alexander von Humboldt. This led to its mining and that of Chilean nitrate and to its application to soil in the United States and Europe after 1840.

The work of Liebig was a revolution for agriculture, and so other investigators started experimentation based on it. In England John Bennet Lawes and Joseph Henry Gilbert worked in the Rothamsted Experimental Station, founded by the former, and discovered that plants took nitrogen from the soil, and that salts needed to be in an available state to be absorbed by plants. Their investigations also produced the "superphosphate", consisting in the acid treatment of phosphate rock. This led to the invention and use of salts of potassium (K) and nitrogen (N) as fertilizers. Ammonia generated by the production of coke was recovered and used as fertiliser. Finally, the chemical basis of nutrients delivered to the soil in manure was understood and in the mid-19th century chemical fertilisers were applied. However, the dynamic interaction of soil and its life forms awaited discovery.

In 1856 J. T. Way discovered that ammonia contained in fertilisers was transformed into nitrates, and twenty years later R. W. Warington proved that this transformation was done by living organisms. In 1890 Sergei Winogradsky announced he had found the bacteria responsible for this transformation.

It was known that certain legumes could take up nitrogen from the air and fix it to the soil but it took the development of bacteriology towards the end of the 19th century to lead to an understanding of the role played in nitrogen fixation by bacteria. The symbiosis of bacteria and leguminous roots, and the fixation of nitrogen by the bacteria, were simultaneously discovered by German agronomist Hermann Hellriegel and Dutch microbiologist Martinus Beijerinck.

Crop rotation, mechanisation, chemical and natural fertilisers led to a doubling of wheat yields in Western Europe between 1800 and 1900.

Studies Concerning Soil Formation

The scientists who studied the soil in connection with agricultural practices had considered it mainly as a static substrate. However, soil is the result of evolution from more ancient geological materials. After studies of the improvement of the soil commenced, others began to study soil genesis and as a result also soil types and classifications.

In 1860, in Mississippi, Eugene W. Hilgard studied the relationship among rock material, climate, and vegetation, and the type of soils that were developed. He realised that the soils were dynamic, and considered soil types classification. Unfortunately his work was not continued. At the same time Vasily Dokuchaev (about 1870) was leading a team of soil scientists in Russia who conducted an extensive survey of soils, finding that similar basic rocks, climate and vegetation types lead to similar soil layering and types, and established the concepts for soil classifications. Due to the language barriers, the work of this team was not communicated to Western Europe until 1914 by a publication in German by K. D. Glinka, a member of the Russian team.

Curtis F. Marbut was influenced by the work of the Russian team, translated Glinka's publication into English, and as he was placed in charge of the U. S. National Cooperative Soil Survey, applied it to a national soil classification system.

Soil-Forming Processes

Soil formation, or pedogenesis, is the combined effect of physical, chemical, biological and anthropogenic processes working on soil parent material. Soil is said to be formed when organic matter has accumulated and colloids are washed downward, leaving deposits of clay, humus, iron oxide, carbonate, and gypsum, producing a distinct layer called the B horizon. This is a somewhat arbitrary definition as mixtures of sand, silt, clay and humus will support biological and agricultural activity before that time. These constituents are moved from one level to another by water and animal activity. As a result, layers (horizons) form in the soil profile. The alteration and movement of materials within a soil causes the formation of distinctive soil horizons.

How soil formation proceeds is influenced by at least five classic factors that are intertwined in the evolution of a soil. They are: parent material, climate, topography (relief), organisms, and time. When reordered to climate, relief, organisms, parent material, and time, they form the acronym CROPT.

An example of the development of a soil would begin with the weathering of lava flow bedrock, which would produce the purely mineral-based parent material from which the soil texture forms. Soil development would proceed most rapidly from bare rock of recent flows in a warm climate, under heavy and frequent rainfall. Under such conditions, plants become established very quickly on basaltic lava, even though there is very little organic material. The plants are supported by the porous rock as it is filled with nutrient-bearing water that carries dissolved minerals from the rocks and guano. Crevasses and pockets, local topography of the rocks, would hold fine materials and harbour plant roots. The developing plant roots are associated with mycorrhizal fungi that assist in breaking up the porous lava, and by these means organic matter and a finer mineral soil accumulate with time.

Parent Material

The mineral material from which a soil forms is called parent material. Rock, whether its origin is igneous, sedimentary, or metamorphic, is the source of all soil mineral materials and the origin of all plant nutrients with the exceptions of nitrogen, hydrogen and carbon. As the parent material is chemically and physically weathered, transported, deposited and precipitated, it is transformed into a soil.

Typical soil parent mineral materials are:

- Quartz: SiO_2
- Calcite: $CaCO_3$
- Feldspar: $KAlSi_3O_8$
- Mica (biotite): $K(Mg,Fe)_3AlSi_3O_{10}(OH)_2$

Classification of Parent Material

Loess field in Germany.

Parent materials are classified according to how they came to be deposited. Residual materials are mineral materials that have weathered in place from primary bedrock. Transported materials are those that have been deposited by water, wind, ice or gravity. Cumulose material is organic matter that has grown and accumulates in place.

Residual soils are soils that develop from their underlying parent rocks and have the same general chemistry as those rocks. The soils found on mesas, plateaux, and plains are residual soils. In the United States as little as three percent of the soils are residual.

Most soils derive from transported materials that have been moved many miles by wind, water, ice and gravity.

- Aeolian processes (movement by wind) are capable of moving silt and fine sand many hundreds of miles, forming loess soils (60–90 percent silt), common in the Midwest of North America and in Central Asia. Clay is seldom moved by wind as it forms stable aggregates.

- Water-transported materials are classed as either alluvial, lacustrine, or marine. Alluvial materials are those moved and deposited by flowing water. Sedimentary deposits settled in lakes are called lacustrine. Lake Bonneville and many soils around the Great Lakes of the United States are examples. Marine deposits, such as soils along the Atlantic and Gulf Coasts and in the Imperial Valley of California of the United States, are the beds of ancient seas that have been revealed as the land uplifted.
- Ice moves parent material and makes deposits in the form of terminal and lateral moraines in the case of stationary glaciers. Retreating glaciers leave smoother ground moraines and in all cases, outwash plains are left as alluvial deposits are moved downstream from the glacier.
- Parent material moved by gravity is obvious at the base of steep slopes as talus cones and is called colluvial material.

Cumulose parent material is not moved but originates from deposited organic material. This includes peat and muck soils and results from preservation of plant residues by the low oxygen content of a high water table. While peat may form sterile soils, muck soils may be very fertile.

Weathering of Parent Material

The weathering of parent material takes the form of physical weathering (disintegration), chemical weathering (decomposition) and chemical transformation. Generally, minerals that are formed under the high temperatures and pressures at great depths within the Earth's mantle are less resistant to weathering, while minerals formed at low temperature and pressure environment of the surface are more resistant to weathering. Weathering is usually confined to the top few meters of geologic material, because physical, chemical, and biological stresses generally decrease with depth. Physical disintegration begins as rocks that have solidified deep in the Earth are exposed to lower pressure near the surface and swell and become mechanically unstable. Chemical decomposition is a function of mineral solubility, the rate of which doubles with each 10 °C rise in temperature, but is strongly dependent on water to effect chemical changes. Rocks that will decompose in a few years in tropical climates will remain unaltered for millennia in deserts. Structural changes are the result of hydration, oxidation, and reduction.

- Physical disintegration is the first stage in the transformation of parent material into soil. Temperature fluctuations cause expansion and contraction of the rock, splitting it along lines of weakness. Water may then enter the cracks and freeze and cause the physical splitting of material along a path toward the center of the rock, while temperature gradients within the rock can cause exfoliation of "shells". Cycles of wetting and drying cause soil particles to be abraded to a finer size, as does the physical rubbing of material as it is moved by wind, water, and gravity. Water can deposit within rocks minerals that expand upon drying, thereby stressing the rock. Finally, organisms reduce parent material in size through the action of plant roots or digging on the part of animals.

- Chemical decomposition and structural changes result when minerals are made soluble by water or are changed in structure. The first three of the following list are solubility changes and the last three are structural changes.

1. The solution of salts in water results from the action of bipolar water on ionic salt compounds producing a solution of ions and water, removing those minerals and reducing the rock's integrity.

2. Hydrolysis is the transformation of minerals into polar molecules by the splitting of intervening water. This results in soluble acid-base pairs. For example, the hydrolysis of orthoclase-feldspar transforms it to acid silicate clay and basic potassium hydroxide, both of which are more soluble.

3. In carbonation, the solution of carbon dioxide in water forms carbonic acid. Carbonic acid will transform calcite into more soluble calcium bicarbonate.

4. Hydration is the inclusion of water in a mineral structure, causing it to swell and leaving it more stressed and easily decomposed.

5. Oxidation of a mineral compound is the inclusion of oxygen in a mineral, causing it to increase its oxidation number and swell due to the relatively large size of oxygen, leaving it stressed and more easily attacked by water (hydrolysis) or carbonic acid (carbonation).

6. Reduction the opposite of oxidation, means the removal of oxygen, hence the oxidation number of some part of the mineral is reduced, which occurs when oxygen is scarce. The reduction of minerals leaves them electrically unstable, more soluble and internally stressed and easily decomposed.

Of the above, hydrolysis and carbonation are the most effective.

Saprolite is a particular example of a residual soil formed from the transformation of granite, metamorphic and other types of bedrock into clay minerals. Often called "weathered granite", saprolite is the result of weathering processes that include: hydrolysis, chelation from organic compounds, hydration (the solution of minerals in water with resulting cation and anion pairs) and physical processes that include freezing and thawing. The mineralogical and chemical composition of the primary bedrock material, its physical features, including grain size and degree of consolidation, and the rate and type of weathering transforms the parent material into a different mineral. The texture, pH and mineral constituents of saprolite are inherited from its parent material.

Climate

The principal climatic variables influencing soil formation are effective precipitation (i.e., precipitation minus evapotranspiration) and temperature, both of which affect the rates of chemical, physical, and biological processes. The temperature and moisture both influence the organic matter content of soil through their effects on the balance between plant growth and microbial decomposition. Climate is the dominant factor in soil formation, and soils show the distinctive characteristics of the climate zones in which they form. For every 10 °C rise in temperature, the rates of biochemical reactions more than double. Mineral precipitation and temperature are the primary climatic influences on soil formation. If warm temperatures and abundant water are present in the profile at the same time, the processes of weathering, leaching, and plant growth will be maximized. Humid climates favor the growth of trees. In contrast, grasses are the dominant native vegetation in subhumid and semiarid regions, while shrubs and brush of various kinds dominate in arid areas.

Water is essential for all the major chemical weathering reactions. To be effective in soil formation, water must penetrate the regolith. The seasonal rainfall distribution, evaporative losses, site topography, and soil permeability interact to determine how effectively precipitation can influence soil formation. The greater the depth of water penetration, the greater the depth of weathering of the soil and its development. Surplus water percolating through the soil profile transports soluble and suspended materials from the upper to the lower layers. It may also carry away soluble materials in the surface drainage waters. Thus, percolating water stimulates weathering reactions and helps differentiate soil horizons. Likewise, a deficiency of water is a major factor in determining the characteristics of soils of dry regions. Soluble salts are not leached from these soils, and in some cases they build up to levels that curtail plant growth. Soil profiles in arid and semi-arid regions are also apt to accumulate carbonates and certain types of expansive clays.

The direct influences of climate include:

- A shallow accumulation of lime in low rainfall areas as caliche
- Formation of acid soils in humid areas
- Erosion of soils on steep hillsides
- Deposition of eroded materials downstream
- Very intense chemical weathering, leaching, and erosion in warm and humid regions where soil does not freeze

Climate directly affects the rate of weathering and leaching. Wind moves sand and smaller particles, especially in arid regions where there is little plant cover. The type and amount of precipitation influence soil formation by affecting the movement of ions and particles through the soil, and aid in the development of different soil profiles. Soil profiles are more distinct in wet and cool climates, where organic materials may accumulate, than in wet and warm climates, where organic materials are rapidly consumed. The effectiveness of water in weathering parent rock material depends on seasonal and daily temperature fluctuations. Cycles of freezing and thawing constitute an effective mechanism which breaks up rocks and other consolidated materials.

Climate also indirectly influences soil formation through the effects of vegetation cover and biological activity, which modify the rates of chemical reactions in the soil.

Topography

The topography, or relief, is characterized by the inclination (slope), elevation, and orientation of the terrain. Topography determines the rate of precipitation or runoff and rate of formation or erosion of the surface soil profile. The topographical setting may either hasten or retard the work of climatic forces.

Steep slopes encourage rapid soil loss by erosion and allow less rainfall to enter the soil before running off and hence, little mineral deposition in lower profiles. In semiarid regions, the lower effective rainfall on steeper slopes also results in less complete vegetative cover, so there is less plant contribution to soil formation. For all of these reasons, steep slopes prevent the formation of soil from getting very far ahead of soil destruction. Therefore, soils on steep terrain tend to have rather shallow, poorly developed profiles in comparison to soils on nearby, more level sites.

In swales and depressions where runoff water tends to concentrate, the regolith is usually more deeply weathered and soil profile development is more advanced. However, in the lowest landscape positions, water may saturate the regolith to such a degree that drainage and aeration are restricted. Here, the weathering of some minerals and the decomposition of organic matter are retarded, while the loss of iron and manganese is accelerated. In such low-lying topography, special profile features characteristic of wetland soils may develop. Depressions allow the accumulation of water, minerals and organic matter and in the extreme, the resulting soils will be saline marshes or peat bogs. Intermediate topography affords the best conditions for the formation of an agriculturally productive soil.

Organisms

Soil is the most abundant ecosystem on Earth, but the vast majority of organisms in soil are microbes, a great many of which have not been described. There may be a population limit of around one billion cells per gram of soil, but estimates of the number of species vary widely from 50,000 per gram to over a million per gram of soil. The total number of organisms and species can vary widely according to soil type, location, and depth.

Plants, animals, fungi, bacteria and humans affect soil formation. Animals, soil mesofauna and micro-organisms mix soils as they form burrows and pores, allowing moisture and gases to move about. In the same way, plant roots open channels in soils. Plants with deep taproots can penetrate many metres through the different soil layers to bring up nutrients from deeper in the profile. Plants with fibrous roots that spread out near the soil surface have roots that are easily decomposed, adding organic matter. Micro-organisms, including fungi and bacteria, effect chemical exchanges between roots and soil and act as a reserve of nutrients.

Humans impact soil formation by removing vegetation cover with erosion as the result. Their tillage also mixes the different soil layers, restarting the soil formation process as less weathered material is mixed with the more developed upper layers.

Earthworms, ants and termites mix the soil as they burrow, significantly affecting soil formation. Earthworms ingest soil particles and organic residues, enhancing the availability of plant nutrients in the material that passes through their bodies. They aerate and stir the soil and increase the stability of soil aggregates, thereby assuring ready infiltration of water. In addition, as ants and termites build mounds, they transport soil materials from one horizon to another.

In general, the mixing of the soil by the activities of animals, sometimes called pedoturbation, tends to undo or counteract the tendency of other soil-forming processes that create distinct horizons. Termites and ants may also retard soil profile development by denuding large areas of soil around their nests, leading to increased loss of soil by erosion. Large animals such as gophers, moles, and prairie dogs bore into the lower soil horizons, bringing materials to the surface. Their tunnels are often open to the surface, encouraging the movement of water and air into the subsurface layers. In localized areas, they enhance mixing of the lower and upper horizons by creating, and later refilling, underground tunnels. Old animal burrows in the lower horizons often become filled with soil material from the overlying A horizon, creating profile features known as crotovinas.

Vegetation impacts soils in numerous ways. It can prevent erosion caused by excessive rain that might result from surface runoff. Plants shade soils, keeping them cooler and slow evaporation of soil moisture, or conversely, by way of transpiration, plants can cause soils to lose moisture. Plants can form new chemicals that can break down minerals and improve the soil structure. The type and amount of vegetation depends on climate, topography, soil characteristics, and biological factors. Soil factors such as density, depth, chemistry, pH, temperature and moisture greatly affect the type of plants that can grow in a given location. Dead plants and fallen leaves and stems begin their decomposition on the surface. There, organisms feed on them and mix the organic material with the upper soil layers; these added organic compounds become part of the soil formation process.

Human activities widely influence soil formation. For example, it is believed that Native Americans regularly set fires to maintain several large areas of prairie grasslands in Indiana and Michigan. In more recent times, human destruction of natural vegetation and subsequent tillage of the soil for crop production has abruptly modified soil formation. Likewise, irrigating an arid region of soil drastically influences the soil-forming factors, as does adding fertilizer and lime to soils of low fertility.

Time

Time is a factor in the interactions of all the above. While a mixture of sand, silt and clay constitute the texture of a soil and the aggregation of those components produces peds, the development of a distinct B horizon marks the development of a soil. With time, soils will evolve features that depend on the interplay of the prior listed soil-forming factors. It takes decades to several thousand years for a soil to develop a profile. That time period depends strongly on climate, parent material, relief, and biotic activity. For example, recently deposited material from a flood exhibits no soil development as there has not been enough time for the material to form a structure that further defines soil. The original soil surface is buried, and the formation process must begin anew for this deposit. Over time the soil will develop a profile that depends on the intensities of biota and climate. While a soil can achieve relative stability of its properties for extended periods, the soil life cycle ultimately ends in soil conditions that leave it vulnerable to erosion. Despite the inevitability of soil retrogression and degradation, most soil cycles are long.

Soil-forming factors continue to affect soils during their existence, even on "stable" landscapes that are long-enduring, some for millions of years. Materials are deposited on top or are blown or washed from the surface. With additions, removals and alterations, soils are always subject to new conditions. Whether these are slow or rapid changes depends on climate, topography and biological activity.

Physical Properties of Soils

The physical properties of soils, in order of decreasing importance, are texture, structure, density, porosity, consistency, temperature, colour and resistivity. Soil texture is determined by the relative proportion of the three kinds of soil mineral particles, called soil separates: sand, silt, and clay. At the next larger scale, soil structures called peds are created from the soil separates when iron oxides, carbonates, clay, silica and humus, coat particles and cause them to adhere into larger, relatively stable secondary structures. Soil density, particularly bulk density, is a measure of soil compaction. Soil porosity consists of the void part of the soil volume and is occupied by gases or water.

Soil consistency is the ability of soil to stick together. Soil temperature and colour are self-defining. Resistivity refers to the resistance to conduction of electric currents and affects the rate of corrosion of metal and concrete structures. These properties may vary through the depth of a soil profile. Most of these properties determine the aeration of the soil and the ability of water to infiltrate and to be held within the soil.

Influence of Soil Texture Separates on Some Properties of Soils			
Property/behavior	**Sand**	**Silt**	**Clay**
Water-holding capacity	Low	Medium to high	High
Aeration	Good	Medium	Poor
Drainage rate	High	Slow to medium	Very slow
Soil organic matter level	Low	Medium to high	High to medium
Decomposition of organic matter	Rapid	Medium	Slow
Warm-up in spring	Rapid	Moderate	Slow
Compactability	Low	Medium	High
Susceptibility to wind erosion	Moderate (High if fine sand)	High	Low
Susceptibility to water erosion	Low (unless fine sand)	High	Low if aggregated, otherwise high
Shrink/Swell Potential	Very Low	Low	Moderate to very high
Sealing of ponds, dams, and landfills	Poor	Poor	Good
Suitability for tillage after rain	Good	Medium	Poor
Pollutant leaching potential	High	Medium	Low (unless cracked)
Ability to store plant nutrients	Poor	Medium to High	High
Resistance to pH change	Low	Medium	High

Texture

Soil types by clay, silt and sand composition as used by the USDA

Iron-rich soil near Paint Pots in Kootenay National Park, Canada

The mineral components of soil are sand, silt and clay, and their relative proportions determine a soil's texture. Properties that are influenced by soil texture, include porosity, permeability, infiltration, shrink-swell rate, water-holding capacity, and susceptibility to erosion. In the illustrated USDA textural classification triangle, the only soil in which neither sand, silt nor clay predominates is called "loam". While even pure sand, silt or clay may be considered a soil, from the perspective of food production a loam soil with a small amount of organic material is considered ideal. The mineral constituents of a loam soil might be 40% sand, 40% silt and the balance 20% clay by weight. Soil texture affects soil behaviour, in particular its retention capacity for nutrients and water.

Sand and silt are the products of physical and chemical weathering of the parent rock; clay, on the other hand, is a most often the product of the precipitation of the dissolved parent rock as a secondary mineral. It is the surface area to volume ratio (specific surface area) of soil particles and the unbalanced ionic charges within those that determine their role in the fertility of soil, as measured by its cation exchange capacity. Sand is least active, followed by silt; clay is the most active. Sand's greatest benefit to soil is that it resists compaction and increases a soil's porosity. Silt is mineralogically like sand but with its higher specific surface area it is more chemically active than sand. But it is the clay content of soil, with its very high specific surface area and generally large number of negative charges, that gives a soil its high retention capacity for water and nutrients. Clay soils also resist wind and water erosion better than silty and sandy soils, as the particles bond tightly to each other.

Sand is the most stable of the mineral components of soil; it consists of rock fragments, primarily quartz particles, ranging in size from 2.0 to 0.05 mm (0.0787 to 0.0020 in) in diameter. Silt ranges in size from 0.05 to 0.002 mm (0.002 to 0.00008 in). Clay cannot be resolved by optical microscopes as its particles are 0.002 mm (7.9×10^{-5} in) or less in diameter and a thickness of only 10 angstroms (10^{-10} m). In medium-textured soils, clay is often washed downward through the soil profile and accumulates in the subsoil.

Soil components larger than 2.0 mm (0.079 in) are classed as rock and gravel and are removed before determining the percentages of the remaining components and the texture class of the soil, but are included in the name. For example, a sandy loam soil with 20% gravel would be called gravelly sandy loam.

When the organic component of a soil is substantial, the soil is called organic soil rather than mineral soil. A soil is called organic if:

1. Mineral fraction is 0% clay and organic matter is 20% or more
2. Mineral fraction is 0% to 50% clay and organic matter is between 20% and 30%
3. Mineral fraction is 50% or more clay and organic matter 30% or more.

Structure

The clumping of the soil textural components of sand, silt and clay causes aggregates to form and the further association of those aggregates into larger units creates soil structures called peds (a contraction of the word pedolith). The adhesion of the soil textural components by organic substances, iron oxides, carbonates, clays, and silica, and the breakage of those aggregates from expansion-contraction, caused by freezing-thawing and wetting-drying cycles, shape soil into distinct geometric forms. The peds evolve into units which have various shapes, sizes and degrees of development. A soil clod, however, is not a ped but rather a mass of soil that results from mechanical disturbance of the soil. Soil structure affects aeration, water movement, conduction of heat, plant root growth and resistance to erosion. Water, in turn, has its strongest effect on soil structure due to its solution and precipitation of minerals and its effect on plant growth.

Soil structure often gives clues to its texture, organic matter content, biological activity, past soil evolution, human use, and the chemical and mineralogical conditions under which the soil formed. While texture is defined by the mineral component of a soil and is an innate property of the soil that does not change with agricultural activities, soil structure can be improved or destroyed by the choice and timing of farming practices.

Soil structural classes:

1. Types: Shape and arrangement of peds
    1. Platy: Peds are flattened one atop the other 1–10 mm thick. Found in the A-horizon of forest soils and lake sedimentation.
    2. Prismatic and Columnar: Prismlike peds are long in the vertical dimension, 10–100 mm wide. Prismatic peds have flat tops, columnar peds have rounded tops. Tend to form in the B-horizon in high sodium soil where clay has accumulated.
    3. Angular and subangular: Blocky peds are imperfect cubes, 5–50 mm, angular have sharp edges, subangular have rounded edges. Tend to form in the B-horizon where clay has accumulated and indicate poor water penetration.
    4. Granular and Crumb: Spheroid peds of polyhedrons, 1–10 mm, often found in the A-horizon in the presence of organic material. Crumb peds are more porous and are considered ideal.
2. Classes: Size of peds whose ranges depend upon the above type
    1. Very fine or very thin: <1 mm platy and spherical; <5 mm blocky; <10 mm prismlike.
    2. Fine or thin: 1–2 mm platy, and spherical; 5–10 mm blocky; 10–20 mm prismlike.
    3. Medium: 2–5 mm platy, granular; 10–20 mm blocky; 20-50 prismlike.
    4. Coarse or thick: 5–10 mm platy, granular; 20–50 mm blocky; 50–100 mm prismlike.
    5. Very coarse or very thick: >10 mm platy, granular; >50 mm blocky; >100 mm prismlike.
3. Grades: Is a measure of the degree of development or cementation within the peds that results in their strength and stability.
    1. Weak: Weak cementation allows peds to fall apart into the three textural constituents, sand, silt and clay.

2. Moderate: Peds are not distinct in undisturbed soil but when removed they break into aggregates, some broken aggregates and little unaggregated material. This is considered ideal.

3. Strong:Peds are distinct before removed from the profile and do not break apart easily.

4. Structureless: Soil is entirely cemented together in one great mass such as slabs of clay or no cementation at all such as with sand.

At the largest scale, the forces that shape a soil's structure result from swelling and shrinkage that initially tend to act horizontally, causing vertically oriented prismatic peds. Clayey soil, due to its differential drying rate with respect to the surface, will induce horizontal cracks, reducing columns to blocky peds. Roots, rodents, worms, and freezing-thawing cycles further break the peds into a spherical shape.

At a smaller scale, plant roots extend into voids and remove water causing the open spaces to increase, thereby decreasing aggregate size. At the same time, roots, fungal hyphae, and earthworms create microscopic tunnels that break up peds.

At an even smaller scale, soil aggregation continues as bacteria and fungi exude sticky polysaccharides which bind soil into smaller peds. The addition of the raw organic matter that bacteria and fungi feed upon encourages the formation of this desirable soil structure.

At the lowest scale, the soil chemistry affects the aggregation or dispersal of soil particles. The clay particles contain polyvalent cations which give the faces of clay layers localized negative charges. At the same time, the edges of the clay plates have a slight positive charge, thereby allowing the edges to adhere to the negative charges on the faces of other clay particles or to flocculate (form clumps). On the other hand, when monovalent ions, such as sodium, invade and displace the polyvalent cations, they weaken the positive charges on the edges, while the negative surface charges are relatively strengthened. This leaves negative charge on the clay faces that repel other clay, causing the particles to push apart, and by doing,the flocculation of clay particles into larger, open assemblages. As a result, the clay disperses and settles into voids between peds, causing those to close. In this way the open structure of the soil is destroyed and the soil is made impenetrable to air and water. Such sodic soil tends to form columnar structures near the surface.

Density

Soil particle density is typically 2.60 to 2.75 grams per cm^3 and is usually unchanging for a given soil. Soil particle density is lower for soils with high organic matter content, and is higher for soils with high iron-oxides content. Soil bulk density is equal to the dry mass of the soil divided by the volume of the soil; i.e., it includes air space and organic materials of the soil volume. The soil bulk density of cultivated loam is about 1.1 to 1.4 g/cm³ (for comparison water is 1.0 g/cm³). Soil bulk density is highly variable for a given soil. A lower bulk density by itself does not indicate suitability for plant growth due to the influence of soil texture and structure. A high bulk density is indicative of either soil compaction or high sand content. Soil bulk density is inherently always less than the soil particle density.

Representative bulk densities of soils. The percentage pore space was calculated using 2.7 g/cm³ for particle density except for the peat soil, which is estimated.		
Soil treatment and identification	**Bulk density g/cm³**	**Pore space %**
Tilled surface soil of a cotton field	1.3	51
Trafficked inter-rows where wheels passed surface	1.67	37
Traffic pan at 25 cm deep	1.7	36
Undisturbed soil below traffic pan, clay loam	1.5	43
Rocky silt loam soil under aspen forest	1.62	40
Loamy sand surface soil	1.5	43
Decomposed peat	0.55	65

Porosity

Pore space is that part of the bulk volume of soil that is not occupied by either mineral or organic matter but is open space occupied by either gases or water. In a productive, medium-textured soil the total pore space is typically about 50% of the soil volume. Pore size varies considerably; the smallest pores (cryptopores; <0.1 μm) hold water too tightly for use by plant roots; plant-available water is held in ultramicropores, micropores and mesopores (0.1-75 μm); and macropores (>75 μm) are generally air-filled when the soil is at field capacity.

Soil texture determines total volume of the smallest pores; clay soils have smaller pores, but more total pore space than sands. Soil structure has a strong influence on the larger pores that affect soil aeration, water infiltration and drainage. Tillage has the short-term benefit of temporarily increasing the number of pores of largest size, but these can be rapidly degraded by the destruction of soil aggregation.

The pore size distribution affects the ability of plants and other organisms to access water and oxygen; large, continuous pores allow rapid transmission of air, water and dissolved nutrients through soil, and small pores store water between rainfall or irrigation events. Pore size variation also compartmentalizes the soil pore space such that many microorganisms are not in direct competition with one another, which may explain not only the large number of species present, but the fact that functionally redundant microorganisms (organisms with the same ecological niche) can co-exist within the same soil.

Consistency

Consistency is the ability of soil to stick to itself or to other objects (cohesion and adhesion respectively) and its ability to resist deformation and rupture. It is of approximate use in predicting cultivation problems and the engineering of foundations. Consistency is measured at three moisture conditions: air-dry, moist, and wet. In those conditions the consistency quality depends upon the clay content. In the wet state, the two qualities of stickiness and plasticity are assessed. A soil's resistance to fragmentation and crumbling is assessed in the dry state by rubbing the sample. Its resistance to shearing forces is assessed in the moist state by thumb and finger pressure. Additionally, the cemented consistency depends on cementation by substances other than clay, such as calcium carbonate, silica, oxides and salts; moisture content has little effect on its assessment. The measures of consistency border on subjective compared to other measures such as pH, since they employ the apparent feel of the soil in those states.

The terms used to describe the soil consistency in three moisture states and a last not affected by the amount of moisture are as follows:

1. Consistency of Dry Soil: loose, soft, slightly hard, hard, very hard, extremely hard
2. Consistency of Moist Soil: loose, very friable, friable, firm, very firm, extremely firm
3. Consistency of Wet Soil: nonsticky, slightly sticky, sticky, very sticky; nonplastic, slightly plastic, plastic, very plastic
4. Consistency of Cemented Soil: weakly cemented, strongly cemented, indurated (requires hammer blows to break up)

Soil consistency is useful in estimating the ability of soil to support buildings and roads. More precise measures of soil strength are often made prior to construction.

Temperature

Soil temperature depends on the ratio of the energy absorbed to that lost. Soil has a temperature range between -20 to 60 °C. Soil temperature regulates seed germination, plant and root growth and the availability of nutrients. Below 50 cm (20 in), soil temperature seldom changes and can be approximated by adding 1.8 °C (2 °F) to the mean annual air temperature. Soil temperature has important seasonal, monthly and daily variations. Fluctuations in soil temperature are much lower with increasing soil depth. Heavy mulching (a type of soil cover) can slow the warming of soil, and, at the same time, reduce fluctuations in surface temperature.

Most often, agricultural activities must adapt to soil temperatures by:

1. maximizing germination and growth by timing of planting

2. optimizing use of anhydrous ammonia by applying to soil below 10 °C (50 °F)
3. preventing heaving and thawing due to frosts from damaging shallow-rooted crops
4. preventing damage to desirable soil structure by freezing of saturated soils
5. improving uptake of phosphorus by plants

Soil temperatures can be raised by drying soils or the use of clear plastic mulches. Organic mulches slow the warming of the soil.

There are various factors that affect soil temperature, such as water content, soil color, and relief (slope, orientation, and elevation), and soil cover (shading and insulation). The color of the ground cover and its insulating properties have a strong influence on soil temperature. Whiter soil tends to have a higher albedo than blacker soil cover, which encourages whiter soils to have lower soil temperatures. The specific heat of soil is the energy required to raise the temperature of soil by 1 °C. The specific heat of soil increases as water content increases, since the heat capacity of water is greater than that of dry soil. The specific heat of pure water is ~ 1 calorie per gram, the specific heat of dry soil is ~ 0.2 calories per gram, hence, the specific heat of wet soil is ~ 0.2 to 1 calories per gram. Also, a tremendous energy (~540 cal/g) is required to evaporate water (known as the heat of vaporization). As such, wet soil usually warms more slowly than dry soil - wet surface soil is typically 3 to 6 °C colder than dry surface soil.

Soil heat flux refers to the rate at which heat energy moves through the soil in response to a temperature difference between two points in the soil. The heat flux density is the amount of energy that flows through soil per unit area per unit time and has both magnitude and direction. For the simple case of conduction into or out of the soil in the vertical direction, which is most often applicable the heat flux density is:

$$q_x = -k\frac{\delta T}{\delta x}$$

In SI units

q is the heat flux density, in SI the units are W·m^{-2}

k is the soils' conductivity, W·m^{-1}·K^{-1}. The thermal conductivity is sometimes a constant, otherwise an average value of conductivity for the soil condition between the surface and the point at depth is used.

δT is the temperature difference (temperature gradient) between the two points in the soil between which the heat flux density is to be calculated. In SI the units are kelvin, K.

δx is the distance between the two points within the soil, at which the temperatures are measured and between which the heat flux density is being calculated. In SI the units are meters m, and where x is measured positive downward.

Heat flux is in the direction opposite the temperature gradient, hence the minus sign. That is to say, if the temperature of the surface is higher than at depth x the negative sign will result in a positive value for the heat flux q, and which is interpreted as the heat being conducted into the soil.

Component	Thermal Conductivity (W·m-1·K-1)
Quartz	8.8
Clay	2.9
Organic matter	0.25
Water	0.57
Ice	2.4
Air	0.025
Dry soil	0.2-0.4
Wet soil	1-3

Soil temperature is important for the survival and early growth of seedlings. Soil temperatures affect the anatomical and morphological character of root systems. All physical, chemical, and biological processes in soil and roots are affected in particular because of the increased viscosities of water and protoplasm at low temperatures. In general, climates that do not preclude survival and growth of white spruce above ground are sufficiently benign to provide soil temperatures able to maintain white spruce root systems. In some northwestern parts of the range, white spruce occurs on permafrost sites and although young unlignified roots of conifers may have little resistance to freezing, less than half of the "secondary mature" root system of white spruce was killed by exposure to a temperature of 23.3 °C in multiple year experiment with containerized trees from local nurseries in Massachusetts.

Optimum temperatures for tree root growth range between 10 °C and 25 °C in general and for spruce in particular. In 2-week-old white spruce seedlings that were then grown for 6 weeks in soil at temperatures of 15 °C, 19 °C, 23 °C, 27 °C, and 31 °C; shoot height, shoot dry weight, stem diameter, root penetration, root volume, and root dry weight all reached maxima at 19 °C.

However, whereas strong positive relationships between soil temperature (5 °C to 25 °C) and growth have been found in trembling aspen and balsam poplar, white and other spruce species have shown little or no changes in growth with increasing soil temperature. Such insensitivity to soil low temperature may be common among a number of western and boreal conifers.

Color

Soil colour is often the first impression one has when viewing soil. Striking colours and contrasting patterns are especially noticeable. The Red River of the South carries sediment eroded from extensive reddish soils like Port Silt Loam in Oklahoma. The Yellow River in China carries yellow sediment from eroding loess soils. Mollisols in the Great Plains of North America are darkened and enriched by organic matter. Podsols in boreal forests have highly contrasting layers due to acidity and leaching.

In general, color is determined by the organic matter content, drainage conditions, and degree of oxidation. Soil color, while easily discerned, has little use in predicting soil characteristics. It is of use in distinguishing boundaries within a soil profile, determining the origin of a soil's parent material, as an indication of wetness and waterlogged conditions, and as a qualitative means of measuring organic, salt and carbonate contents of soils. Color is recorded in the Munsell color system as for instance 10YR3/4 *Dusky Red.*

Soil color is primarily influenced by soil mineralogy. Many soil colours are due to various iron minerals. The development and distribution of colour in a soil profile result from chemical and biological weathering, especially redox reactions. As the primary minerals in soil parent material weather, the elements combine into new and colourful compounds. Iron forms secondary minerals of a yellow or red colour, organic matter decomposes into black and brown compounds, and manganese, sulfur and nitrogen can form black mineral deposits. These pigments can produce various colour patterns within a soil. Aerobic conditions produce uniform or gradual colour changes, while reducing environments (anaerobic) result in rapid colour flow with complex, mottled patterns and points of colour concentration.

Resistivity

Soil resistivity is a measure of a soil's ability to retard the conduction of an electric current. The electrical resistivity of soil can affect the rate of galvanic corrosion of metallic structures in contact with the soil. Higher moisture content or increased electrolyte concentration can lower resistivity and increase conductivity, thereby increasing the rate of corrosion. Soil resistivity values typically range from about 2 to 1000 $\Omega \cdot m$, but more extreme values are not unusual.

Soil Water

Water that enters a field is removed from a field by runoff, drainage, evaporation or transpiration. Runoff is the water that flows on the surface to the edge of the field; drainage is the water that flows through the soil downward or toward the edge of the field underground; evaporative water loss from a field is that part of the water that evaporates into the atmosphere directly from the field's surface; transpiration is the loss of water from the field by its evaporation from the plant itself.

Water affects soil formation, structure, stability and erosion but is of primary concern with respect to plant growth. Water is essential to plants for four reasons:

1. It constitutes 80%-95% of the plant's protoplasm.
2. It is essential for photosynthesis.
3. It is the solvent in which nutrients are carried to, into and throughout the plant.
4. It provides the turgidity by which the plant keeps itself in proper position.

In addition, water alters the soil profile by dissolving and re-depositing minerals, often at lower levels, and possibly leaving the soil sterile in the case of extreme rainfall and drainage. In a loam soil, solids constitute half the volume, gas one-quarter of the volume, and water one-quarter of the volume of which only half of which will be available to most plants.

A flooded field will drain the gravitational water under the influence of gravity until water's adhesive and cohesive forces resist further drainage at which point it is said to have reached field capacity. At that point, plants must apply suction to draw water from a soil. When soil becomes too dry, the available water is used up and the remaining moisture is called unavailable water as the plant cannot produce sufficient suction to draw that water in. A plant must produce suction that increases from zero for a flooded field to 1/3 bar at field dry condition (one bar is a little less than one atmosphere pressure). At 15 bar suction, wilting percent, seeds will not germinate, plants begin to wilt and then die. Water moves in soil under the influence of gravity, osmosis and capillarity. When water enters the soil, it displaces air from some of the pores, since air content of a soil is inversely related to its water content.

The rate at which a soil can absorb water depends on the soil and its other conditions. As a plant grows, its roots remove water from the largest pores first. Soon the larger pores hold only air, and the remaining water is found only in the intermediate- and smallest-sized pores. The water in the smallest pores is so strongly held to particle surfaces that plant roots cannot pull it away. Consequently, not all soil water is available to plants. When saturated, the soil may lose nutrients as the water drains. Water moves in a drained field under the influence of pressure where the soil is locally saturated and by capillarity pull to dryer parts of the soil. Most plant water needs are supplied from the suction caused by evaporation from plant leaves and 10% is supplied by "suction" created by osmotic pressure differences between the plant interior and the soil water. Plant roots must seek out water. Insufficient water will damage the yield of a crop. Most of the available water is used in transpiration to pull nutrients into the plant.

Water Retention Forces

Water is retained in a soil when the adhesive force of attraction that water's hydrogen atoms have for the oxygen of soil particles is stronger than the cohesive forces that water's hydrogen feels for other water oxygen atoms. When a field is flooded, the soil pore space is completely filled by water. The field will drain under the force of gravity until it reaches what is called field capacity, at which point the smallest pores are filled with water and the largest with water and gases. The total amount of water held when field capacity is reached is a function of the specific surface area of the soil particles. As a result, high clay and high organic soils have higher field capacities. The total force required to pull or push water out of soil is termed suction and usually expressed in units of bars (10^5 pascal) which is just a little less than one-atmosphere pressure. Alternatively, the terms "tension" or "moisture potential" may be used.

Moisture Classification

The forces with which water is held in soils determine its availability to plants. Forces of adhesion hold water strongly to mineral and humus surfaces and less strongly to itself by cohesive forces. A plant's root may penetrate a very small volume of water that is adhering to soil and be initially able to draw in water that is only lightly held by the cohesive forces. But as the droplet is drawn down, the forces of adhesion of the water for the soil particles make reducing the volume of water increasingly difficult until the plant cannot produce sufficient suction to use the remaining water. The remaining water is considered unavailable. The amount of available water depends upon the soil texture and humus amounts and the type of plant attempting to use the water. Cacti, for example, can produce greater suction than can agricultural crop plants.

The following description applies to a loam soil and agricultural crops. When a field is flooded, it is said to be saturated and all available air space is occupied by water. The suction required to draw water into a plant root is zero. As the field drains under the influence of gravity (drained water is called gravitational water or drain-able water), the suction a plant must produce to use such water increases to 1/3 bar. At that point, the soil is said to have reached field capacity, and plants that use the water must produce increasingly higher suction, finally up to 15 bar. At 15 bar suction, the soil water amount is called wilting percent. At that suction the plant cannot sustain its water needs as water is still being lost from the plant by transpiration; the plant's turgidity is lost, and it wilts. The next level, called air-dry, occurs at 1000 bar suction. Finally the oven dry condition is reached at 10,000 bar suction. All water below wilting percentage is called unavailable water.

Soil Moisture Content

When the soil moisture content is optimal for plant growth, the water in the large and intermediate size pores can move about in the soil and be easily used by plants. The amount of water remaining in a soil drained to field capacity and the amount that is available are functions of the soil type. Sandy soil will retain very little water, while clay will hold the maximum amount. The time required to drain a field from flooded condition for a clay loam that begins at 43% water by weight to a field capacity of 21.5% is six days, whereas a sand loam that is flooded to its maximum of 22% water will take two days to reach field capacity of 11.3% water. The available water for the clay loam might be 11.3% whereas for the sand loam it might be only 7.9% by weight.

Wilting point, field capacity, and available water capacity of various soil textures						
Soil Texture	Wilting Point		Field Capacity		Available water capacity	
	Water per foot of soil depth		Water per foot of soil depth		Water per foot of soil depth	
	%	in.	%	in.	%	in.
Medium sand	1.7	0.3	6.8	1.2	5.1	0.9
Fine sand	2.3	0.4	8.5	1.5	6.2	1.1
Sandy loam	3.4	0.6	11.3	2.0	7.9	1.4
Fine sandy loam	4.5	0.8	14.7	2.6	10.2	1.8
Loam	6.8	1.2	18.1	3.2	11.3	2.0
Silt loam	7.9	1.4	19.8	3.5	11.9	2.1
Clay loam	10.2	1.8	21.5	3.8	11.3	2.0
Clay	14.7	2.6	22.6	4.0	7.9	1.4

The above are average values for the soil textures as the percentage of sand, silt and clay vary within the listed soil textures.

Water Flow in Soils

Water moves through soil due to the force of gravity, osmosis and capillarity. At zero to one-third bar suction, water is pushed through soil from the point of its application under the force of gravity and the pressure gradient created by the pressure of the water; this is called saturated flow. At higher suction, water movement is pulled by capillarity from wetter toward dryer soil. This is caused by water's adhesion to soil solids, and is called unsaturated flow.

Water infiltration and movement in soil is controlled by six factors:

1. Soil texture
2. Soil structure. Fine-textured soils with granular structure are most favourable to infiltration of water.
3. The amount of organic matter. Coarse matter is best and if on the surface helps prevent the destruction of soil structure and the creation of crusts.
4. Depth of soil to impervious layers such as hardpans or bedrock
5. The amount of water already in the soil
6. Soil temperature. Warm soils take in water faster while frozen soils may not be able to absorb depending on the type of freezing.

Water infiltration rates range from 0.25 cm (0.098 in) per hour for high clay soils to 2.5 cm (0.98 in) per hour for sand and well stabilised and aggregated soil structures. Water flows through the ground unevenly, called "gravity fingers", because of the surface tension between water particles. Tree roots create paths for rainwater flow through soil by breaking though soil including clay layers: one study showed roots increasing infiltration of water by 153% and another study showed an increase by 27 times. Flooding temporarily increases soil permeability in river beds, helping to recharge aquifers.

Saturated Flow

Water applied to a soil is pushed by pressure gradients from the point of its application where it is saturated locally, to less saturated areas. Once soil is completely wetted, any more water will move downward, or percolate, carrying with it clay, humus and nutrients, primarily cations, out of the range of plant roots. In order of decreasing solubility, the leached nutrients are:

- Calcium
- Magnesium, Sulfur, Potassium; depending upon soil composition
- Nitrogen; usually little, unless nitrate fertiliser was applied recently
- Phosphorus; very little as its forms in soil are of low solubility.

In the United States percolation water due to rainfall ranges from zero inches just east of the Rocky Mountains to twenty or more inches in the Appalachian Mountains and the north coast of the Gulf of Mexico.

Unsaturated Flow

At suctions less than one-third bar, water moves in all directions via unsaturated flow at a rate that is dependent on the square of the diameter of the water-filled pores. Water is pulled by capillary action due to the adhesion force of water to the soil solids, producing a suction gradient from wet towards drier soil. Doubling the diameter of the pores increases the flow rate by a factor of four. Large pores drained by gravity and not filled with water do not greatly increase the flow rate for unsaturated flow. Water flow is primarily from coarse-textured soil into fine-textured soil and is slowest in fine-textured soils such as clay.

Water Uptake by Plants

Of equal importance to the storage and movement of water in soil is the means by which plants acquire it and their nutrients. Ninety percent of water is taken up by plants as passive absorption caused by the pulling force of water evaporating (transpiring) from the long column of water that leads from the plant's roots to its leaves. In addition, the high concentration of salts within plant roots creates an osmotic pressure gradient that pushes soil water into the roots. Osmotic absorption becomes more important during times of low water transpiration caused by lower temperatures (for example at night) or high humidity. It is the process that causes guttation.

Root extension is vital for plant survival. A study of a single winter rye plant grown for four months in one cubic foot of loam soil showed that the plant developed 13,800,000 roots, a total of 385 miles in length with 2,550 square feet in surface area; and 14 billion hair roots of 6,600 miles total length and 4,320 square feet total area; for a total surface area of 6,870 square feet (83 ft squared). The total surface area of the loam soil was estimated to be 560,000 square feet. In other words, the roots were in contact with only 1.2% of the soil.

Roots must seek out water as the unsaturated flow of water in soil can move only at a rate of up to 2.5 cm (one inch) per day; as a result they are constantly dying and growing as they seek out high concentrations of soil moisture. Insufficient soil moisture, to the point of causing wilting, will cause permanent damage and crop yields will suffer. When grain sorghum was exposed to soil suction as low as 13.0 bar during the seed head emergence through bloom and seed set stages of growth, its production was reduced by 34%.

Consumptive Use and Water Efficiency

Only a small fraction (0.1% to 1%) of the water used by a plant is held within the plant. The majority is ultimately lost via transpiration, while evaporation from the soil surface is also substantial. Transpiration plus evaporative soil moisture loss is called evapotranspiration. Evapotranspiration plus water held in the plant totals to consumptive use, which is nearly identical to evapotranspiration.

The total water used in an agricultural field includes runoff, drainage and consumptive use. The use of loose mulches will reduce evaporative losses for a period after a field is irrigated, but in the end the total evaporative loss will approach that of an uncovered soil. The benefit from mulch is to keep the moisture available during the seedling stage. Water use efficiency is measured by transpiration ratio, which is the ratio of the total water transpired by a plant to the dry weight of the harvested plant. Transpiration ratios for crops range from 300 to 700. For example, alfalfa may have a transpiration ratio of 500 and as a result 500 kilograms of water will produce one kilogram of dry alfalfa.

Soil Atmosphere

The atmosphere of soil is radically different from the atmosphere above. The consumption of oxygen, by microbes and plant roots and their release of carbon dioxide, decrease oxygen and increase carbon dioxide concentration. Atmospheric CO_2 concentration is 0.04%, but in the soil pore space it may range from 10 to 100 times that level. At extreme levels CO_2 is toxic. In addition, the soil voids are saturated with water vapour. Adequate porosity is necessary, not just to allow the penetration of water, but also to allow gases to diffuse in and out. Movement of gases is by diffusion from high concentrations to lower. Oxygen diffuses in and is consumed and excess levels of carbon dioxide, diffuse out with other gases as well as water. Soil texture and structure strongly affect soil porosity and gas diffusion. It is the total pore space (porosity) of soil not the pore size that determines the rate of diffusion of gases into and out of soil. A Platy soil structure and compacted soils (low porosity) impede gas flow, and a deficiency of oxygen may encourage anaerobic bacteria to reduce (strip oxygen) from nitrate NO_3 to the gases N_2, N_2O, and NO, which are then lost to the atmosphere, thereby depleting the soil of nitrogen. Aerated soil is also a net sink of methane CH_4 but a net producer of greenhouse gases when soils are depleted of oxygen and subject to elevated temperatures.

Composition of Soil Particles

Soil particles can be classified by their chemical composition (mineralogy) as well as their size. The particle size distribution of a soil, its texture, determines many of the properties of that soil, but the mineralogy of those particles can strongly modify those properties. The mineralogy of the finest soil particles, clay, is especially important.

Gravel, Sand and Silt

Gravel, sand and silt are the larger soil particles, and their mineralogy is often inherited from the parent material of the soil, but may include products of weathering (such as concretions of calcium carbonate or iron oxide), or residues of plant and animal life (such as silica phytoliths). Quartz is the most common mineral in the sand or silt fraction as it is resistant to chemical weathering; other common minerals are feldspars, micas and ferromagnesian minerals such as pyroxenes, amphiboles and olivines.

Mineral Colloids; Soil Clays

Due to its high specific surface area and its unbalanced negative charges, clay is the most active mineral component of soil. It is a colloidal and most often a crystalline material. In soils, clay is a soil textural class and is defined in a physical sense as any mineral particle less than 2 μm (8×10^{-5} in) in effective diameter. Many soil minerals, such as gypsum, carbonates, or quartz, are small enough to be classified as clay based on their physical size, but chemically they do not afford the same utility as do clay minerals. Chemically, clay is a range of minerals with certain reactive properties.

Clay was once thought to be very small particles of quartz, feldspar, mica, hornblende or augite, but it is now known to be (with the exception of mica-based clays) a precipitate with a mineralogical composition that is dependent on but different from its parent materials and is classed as a secondary mineral. The type of clay that is formed is a function of the parent material and the composition of the minerals in solution. Clay minerals continue to be formed as long as the soil exists. Mica-based clays result from a modification of the primary mica mineral in such a way that it behaves and is classed as a clay. Most clays are crystalline, but some are amorphous. The clays of a soil are a mixture of the various types of clay, but one type predominates.

There are four groups of clay: layer silicates; crystalline chain silicates; metal oxides and hydroxides and oxy-oxides; and amorphous; and allophanes. Most clays are crystalline and most are made up of three or four planes of oxygen held together by planes of aluminium and silicon by way of ionic bonds that together form a single layer of clay. The spatial arrangement of the oxygen atoms determines clay's structure. Half of the weight of clay is oxygen, but on a volume basis oxygen is ninety percent. The layers of clay are sometimes held together through hydrogen bonds or potassium bridges and as a result will swell less in the presence of water. Other clays, such as montmorillonite, have layers that are loosely attached and will swell greatly when water intervenes between the layers.

There are four groups of clays:

1. Layer Crystalline alumino-silica clays: montmorillonite, illite, vermiculite, chlorite, kaolinite.

2. Crystalline Chain carbonate and sulfate minerals: calcite ($CaCO_3$), dolomite ($CaMg(CO_3)_2$) and gypsum ($CaSO_4 \cdot 2H2O$).

3. Amorphous clays: young mixtures of silica (SiO_2-OH) and alumina ($Al(OH)_3$) which have not had time to form regular crystals.

4. Sesquioxide clays: old, highly leached clays which result in oxides of iron, aluminium and titanium.

Alumino-Silica Clays

Alumino-silica clays are characterised by their regular crystalline structure. Oxygen in ionic bonds with silicon forms a tetrahedral coordination (silicon at the center) which in turn forms sheets of silica. Two sheets of silica are bonded together by a plane of aluminium which forms an octahedral coordination, called alumina, with the oxygens of the silica sheet above and that below it. Hydroxyl ions (OH^-) sometimes substitute for oxygen. During the clay formation process, Al^{3+} may substitute for Si^{4+} in the silica layer, and as much as one fourth of the aluminium Al^{3+} may be substituted by Zn^{2+},

Mg^{2+} or Fe^{2+} in the alumina layer. The substitution of lower-valence cations for higher-valence cations (isomorphous substitution) gives clay a local negative charge on an oxygen atom that attracts and holds water and positively charged soil cations, some of which are of value for plant growth. Isomorphous substitution occurs during the clay's formation and does not change with time.

- Montmorillonite clay is made of four planes of oxygen with two silicon and one central aluminium plane intervening. The alumino-silicate montmorillonite clay is said to have a 2:1 ratio of silicon to aluminium. The seven planes together form a single crystal of montmorillonite. The crystals are weakly held together and water may intervene, causing the clay to swell up to ten times its dry volume. It occurs in soils which have had little leaching, hence it is found in arid regions. As the crystals are not bonded face to face, the entire surface is exposed and available for surface reactions, hence it has a high cation exchange capacity (CEC).

- Illite is a 2:1 clay similar in structure to montmorillonite but has potassium bridges between the faces of the clay crystals and the degree of swelling depends on the degree of weathering of the potassium. The active surface area is reduced due to the potassium bonds. Illite originates from the modification of mica, a primary mineral. It is often found together with montmorillonite and its primary minerals. It has moderate CEC.

- Vermiculite is a mica-based clay similar to illite, but the crystals of clay are held together more loosely by hydrated magnesium and it will swell, but not as much as does montmorillonite. It has very high CEC.

- Chlorite is similar to vermiculite, but the loose bonding by occasional hydrated magnesium, as in vermiculite, is replaced by a hydrated magnesium sheet, that firmly bonds the planes above and below it. It has two planes of silicon, one of aluminium and one of magnesium; hence it is a 2:2 clay. Chlorite does not swell and it has low CEC.

- Kaolinite is very common, highly weathered clay, and more common than montmorillonite in acid soils. It has one silica and one alumina plane per crystal; hence it is a 1:1 type clay. One plane of silica of montmorillonite is dissolved and is replaced with hydroxyls, which produces strong hydrogen bonds to the oxygen in the next crystal of clay. As a result, kaolinite does not swell in water and has a low specific surface area, and as almost no isomorphous substitution has occurred it has a low CEC. Where rainfall is high, acid soils selectively leach more silica than alumina from the original clays, leaving kaolinite. Even heavier weathering results in sesquioxide clays.

silica-sesquioxide

Crystalline Chain Clays

The carbonate and sulfate minerals are much more soluble and hence are found primarily in desert soils where leaching is less active.

Amorphous Clays

Amorphous clays are young, and commonly found in volcanic ash. They are mixtures of alumina and silica which have not formed the ordered crystal shape of alumino-silica clays which time would provide. The majority of their negative charges originates from hydroxyl ions, which can gain or lose a hydrogen ion (H^+) in response to soil pH, in such way was as to buffer the soil pH. They may have either a negative charge provided by the attached hydroxyl ion (OH^-), which can attract a cation, or lose the hydrogen of the hydroxyl to solution and display a positive charge which can attract anions. As a result, they may display either high CEC in an acid soil solution, or high anion exchange capacity in a basic soil solution.

Sesquioxide Clays

Sesquioxide clays are a product of heavy rainfall that has leached most of the silica from alumino-silica clay, leaving the less soluble oxides iron hematite (Fe_2O_3), iron hydroxide ($Fe(OH)_3$), aluminium hydroxide gibbsite ($Al(OH)_3$), hydrated manganese birnessite (MnO_2). It takes hundreds of thousands of years of leaching to create sesquioxide clays. *Sesqui* is Latin for "one and one-half": there are three parts oxygen to two parts iron or aluminium; hence the ratio is one and one-half (not true for all). They are hydrated and act as either amorphous or crystalline. They are not sticky and do not swell, and soils high in them behave much like sand and can rapidly pass water. They are able to hold large quantities of phosphates. Sesquioxides have low CEC but are able to hold anions as well as cations. Such soils range from yellow to red in colour. Such clays tend to hold phosphorus so tightly that it is unavailable for absorption by plants.

Organic Colloids

Humus is the final state of decomposition of organic matter. While it may linger for a thousand years, on the larger scale of the age of the mineral soil components, it is temporary. It is composed of the very stable lignins (30%) and complex sugars (polyuronides, 30%), proteins (30%), waxes, and fats that are resistant to breakdown by microbes. Its chemical assay is 60% carbon, 5% nitrogen, some oxygen and the remainder hydrogen, sulfur, and phosphorus. On a dry weight basis, the CEC of humus is many times greater than that of clay.

Carbon and Terra Preta

In the extreme environment of high temperatures and the leaching caused by the heavy rain of tropical rain forests, the clay and organic colloids are largely destroyed. The heavy rains wash the alumino-silicate clays from the soil leaving only sesquioxide clays of low CEC. The high temperatures and humidity allow bacteria and fungi to virtually dissolve any organic matter on the rain-forest floor overnight and much of the nutrients are volatilized or leached from the soil and lost. However, carbon in the form of charcoal is far more stable than soil colloids and is capable of performing many of the functions of the soil colloids of sub-tropical soils. Soil containing substantial quantities of charcoal, of an anthropogenic origin, is called terra preta. Research into terra preta is still young but is promising. Fallow periods "on the Amazonian Dark Earths can be as short as 6 months, whereas fallow periods on oxisols are usually 8 to 10 years long"

Soil Chemistry

The chemistry of a soil determines its ability to supply available plant nutrients and affects its physical properties and the health of its microbial population. In addition, a soil's chemistry also determines its corrosivity, stability, and ability to absorb pollutants and to filter water. It is the surface chemistry of mineral and organic colloids that determines soil's chemical properties. "A colloid is a small, insoluble, nondiffusible particle larger than a molecule but small enough to remain suspended in a fluid medium without settling. Most soils contain organic colloidal particles called humus as well as the inorganic colloidal particles of clays." The very high specific surface area of colloids and their net charges, gives soil its ability to hold and release ions. Negatively charged sites on colloids attract and release cations in what is referred to as cation exchange. Cation-exchange capacity (CEC) is the amount of exchangeable cations per unit weight of dry soil and is expressed in terms of milliequivalents of positively charged ions per 100 grams of soil (or centimoles of positive charge per kilogram of soil; $cmol_c/kg$). Similarly, positively charged sites on colloids can attract and release anions in the soil giving the soil anion exchange capacity (AEC).

Cation and Anion Exchange

The cation exchange, that takes place between colloids and soil water, buffers (moderates) soil pH, alters soil structure, and purifies percolating water by adsorbing cations of all types, both useful and harmful.

The negative or positive charges on colloid particles make them able to hold cations or anions, respectively, to their surfaces. The charges result from four sources.

1. Isomorphous substitution occurs in clay during its formation, when lower-valence cations substitute for higher-valence cations in the crystal structure. Substitutions in the outermost layers are more effective than for the innermost layers, as the charge strength drops off as the square of the distance. The net result is oxygen atoms with net negative charge and the ability to attract cations.
2. Edge-of-clay oxygen atoms are not in balance ionically as the tetrahedral and octahedral structures are incomplete.
3. Hydroxyls may substitute for oxygens of the silica layers. When the hydrogens of the clay hydroxyls are ionised into solution, they leave the oxygen with a negative charge.
4. Hydrogens of humus hydroxyl groups may be ionised into solution, leaving an oxygen with a negative charge.

Cations held to the negatively charged colloids resist being washed downward by water and out of reach of plants' roots, thereby preserving the fertility of soils in areas of moderate rainfall and low temperatures.

There is a hierarchy in the process of cation exchange on colloids, as they differ in the strength of adsorption by the colloid and hence their ability to replace one another. If present in equal amounts in the soil water solution:

Al^{3+} replaces H^+ replaces Ca^{2+} replaces Mg^{2+} replaces K^+ same as NH^{4+} replaces Na^+

If one cation is added in large amounts, it may replace the others by the sheer force of its numbers. This is called mass action. This is largely what occurs with the addition of fertiliser.

As the soil solution becomes more acidic (low pH, and an abundance of H^+), the other cations more weakly bound to colloids are pushed into solution as hydrogen ions occupy those sites. A low pH may cause hydrogen of hydroxyl groups to be pulled into solution, leaving charged sites on the colloid available to be occupied by other cations. This ionisation of hydroxyl groups on the surface of soil colloids creates what is described as pH-dependent charges. Unlike permanent charges developed by isomorphous substitution, pH-dependent charges are variable and increase with increasing pH. Freed cations can be made available to plants but are also prone to be leached from the soil, possibly making the soil less fertile. Plants are able to excrete H^+ into the soil and by that means, change the pH of the soil near the root and push cations off the colloids, thus making those available to the plant.

Cation Exchange Capacity (CEC)

Cation exchange capacity should be thought of as the soil's ability to remove cations from the soil water solution and sequester those to be exchanged later as the plant roots release hydrogen ions to the solution. CEC is the amount of exchangeable hydrogen cation (H^+) that will combine with 100 grams dry weight of soil and whose measure is one milliequivalents per 100 grams of soil (1 meq/100 g). Hydrogen ions have a single charge and one-thousandth of a gram of hydrogen ions per 100 grams dry soil gives a measure of one milliequivalent of hydrogen ion. Calcium, with an atomic weight 40 times that of hydrogen and with a valence of two, converts to (40/2) x 1 milliequivalent = 20 milliequivalents of hydrogen ion per 100 grams of dry soil or 20 meq/100 g. The modern measure of CEC is expressed as centimoles of positive charge per kilogram (cmol/kg) of oven-dry soil.

Most of the soil's CEC occurs on clay and humus colloids, and the lack of those in hot, humid, wet climates, due to leaching and decomposition respectively, explains the relative sterility of tropical soils. Live plant roots also have some CEC.

Cation exchange capacity for soils; soil textures; soil colloids		
Soil	**State**	**CEC meq/100 g**
Charlotte fine sand	Florida	1.0
Ruston fine sandy loam	Texas	1.9
Glouchester loam	New Jersey	11.9
Grundy silt loam	Illinois	26.3
Gleason clay loam	California	31.6
Susquehanna clay loam	Alabama	34.3
Davie mucky fine sand	Florida	100.8
Sands	------	1 - 5
Fine sandy loams	------	5-10
Loams and silt loams	-----	5-15
Clay loams	-----	15-30
Clays	-----	over 30

Sesquioxides	-----	0-3
Kaolinite	-----	3-15
Illite	-----	25-40
Montmorillonite	-----	60-100
Vermiculite (similar to illite)	-----	80-150
Humus	-----	100-300

Anion Exchange Capacity (AEC)

Anion exchange capacity should be thought of as the soil's ability to remove anions from the soil water solution and sequester those for later exchange as the plant roots release carbonate anions to the soil water solution. Those colloids which have low CEC tend to have some AEC. Amorphous and sesquioxide clays have the highest AEC, followed by the iron oxides. Levels of AEC are much lower than for CEC. Phosphates tend to be held at anion exchange sites.

Iron and aluminum hydroxide clays are able to exchange their hydroxide anions (OH^-) for other anions. The order reflecting the strength of anion adhesion is as follows:

$$H_2PO_4^- \text{ replaces } SO_4^{2-} \text{ replaces } NO_3^- \text{ replaces } Cl^-$$

The amount of exchangeable anions is of a magnitude of tenths to a few milliequivalents per 100 g dry soil. As pH rises, there are relatively more hydroxyls, which will displace anions from the colloids and force them into solution and out of storage; hence AEC decreases with increasing pH (alkalinity).

Soil Reaction (pH)

Soil reactivity is expressed in terms of pH and is a measure of the acidity or alkalinity of the soil. More precisely, it is a measure of hydrogen ion concentration in an aqueous solution and ranges in values from 0 to 14 (acidic to basic) but practically speaking for soils, pH ranges from 3.5 to 9.5, as pH values beyond those extremes are toxic to life forms.

Soil pH

At 25 °C an aqueous solution that has a pH of 3.5 has $10^{-3.5}$ moles H^+ (hydrogen ions) per litre of solution (and also $10^{-10.5}$ mole/litre OH^-). A pH of 7, defined as neutral, has 10^{-7} moles hydrogen ions per litre of solution and also 10^{-7} moles of OH^- per litre; since the two concentrations are equal, they are said to neutralise each other. A pH of 9.5 has $10^{-9.5}$ moles hydrogen ions per litre of solution (and also $10^{-2.5}$ mole per litre OH^-). A pH of 3.5 has one million times more hydrogen ions per litre than a solution with pH of 9.5 ($9.5 - 3.5 = 6$ or 10^6) and is more acidic.

The effect of pH on a soil is to remove from the soil or to make available certain ions. Soils with high acidity tend to have toxic amounts of aluminium and manganese.

Plants which need calcium need moderate alkalinity, but most minerals are more soluble in acid soils. Soil organisms are hindered by high acidity, and most agricultural crops do best with mineral soils of pH 6.5 and organic soils of pH 5.5.

In high rainfall areas, soils tend to acidity as the basic cations are forced off the soil colloids by the mass action of hydrogen ions from the rain as those attach to the colloids. High rainfall rates can then wash the nutrients out, leaving the soil sterile. Once the colloids are saturated with H^+, the addition of any more hydrogen ions or aluminum hydroxyl cations drives the pH even lower (more acidic) as the soil has been left with no buffering capacity. In areas of extreme rainfall and high temperatures, the clay and humus may be washed out, further reducing the buffering capacity of the soil. In low rainfall areas, unleached calcium pushes pH to 8.5 and with the addition of exchangeable sodium, soils may reach pH 10. Beyond a pH of 9, plant growth is reduced. High pH results in low micro-nutrient mobility, but water-soluble chelates of those nutrients can correct the deficit. Sodium can be reduced by the addition of gypsum (calcium sulphate) as calcium adheres to clay more tightly than does sodium causing sodium to be pushed into the soil water solution where it can be washed out by an abundance of water.

Base Saturation Percentage

There are acid-forming cations (hydrogen and aluminium) and there are base-forming cations. The fraction of the base-forming cations that occupy positions on the soil colloids is called the base saturation percentage. If a soil has a CEC of 20 meq and 5 meq are aluminium and hydrogen cations (acid-forming), the remainder of positions on the colloids (20-5 = 15 meq) are assumed occupied by base-forming cations, so that the percentage base saturation is 15/20 x 100% = 75% (the compliment 25% is assumed acid-forming cations). When the soil pH is 7 (neutral), base saturation is 100 percent and there are no hydrogen ions stored on the colloids. Base saturation is almost in direct proportion to pH (increases with increasing pH). It is of use in calculating the amount of lime needed to neutralise an acid soil. The amount of lime needed to neutralize a soil must take account of the amount of acid forming ions on the colloids not just those in the soil water solution. The addition of enough lime to neutralize the soil water solution will be insufficient to change the pH, as the acid forming cations stored on the soil colloids will tend to restore the original pH condition as they are pushed off those colloids by the calcium of the added lime.

Buffering of Soils

The resistance of soil to changes in pH, as a result of the addition of acid or basic material, is a measure of the buffering capacity of a soil and increases as the CEC increases. Hence, pure sand has almost no buffering ability, while soils high in colloids have high buffering capacity. Buffering occurs by cation exchange and neutralisation.

The addition of a small amount highly basic aqueous ammonia to a soil will cause the ammonium to displace hydrogen ions from the colloids, and the end product is water and colloidally fixed ammonium, but no permanent change overall in soil pH.

The addition of a small amount of lime, $CaCO_3$, will displace hydrogen ions from the soil colloids, causing the fixation of calcium to colloids and the evolution of CO_2 and water, with no permanent change in soil pH.

The addition of carbonic acid (the solution of CO_2 in water) will displace calcium from colloids, as hydrogen ions are fixed to the colloids, evolving water and slightly alkaline (temporary increase in pH) highly soluble calcium bicarbonate, which will then precipitate as lime ($CaCO_3$) and water at a lower level in the soil profile, with the result of no permanent change in soil pH.

All of the above are examples of the buffering of soil pH. The general principal is that an increase in a particular cation in the soil water solution will cause that cation to be fixed to colloids (buffered) and a decrease in solution of that cation will cause it to be withdrawn from the colloid and moved into solution (buffered). The degree of buffering is limited by the CEC of the soil; the greater the CEC, the greater the buffering capacity of the soil.

Nutrients

Sixteen elements or nutrients are essential for plant growth and reproduction. They are carbon C, hydrogen H, oxygen O, nitrogen N, phosphorus P, potassium K, sulfur S, calcium Ca, magnesium Mg, iron Fe, boron B, manganese Mn, copper Cu, zinc Zn, molybdenum Mo, and chlorine Cl. Nutrients required for plants to complete their life cycle are considered essential nutrients. Nutrients that enhance the growth of plants but are not necessary to complete the plant's life cycle are considered non-essential. With the exception of carbon, hydrogen and oxygen, which are supplied by carbon dioxide and water, the nutrients derive originally from the mineral component of the soil.

Plant uptake of nutrients can only proceed when they are present in a plant-available form. In most situations, nutrients are absorbed in an ionic form from (or together with) soil water. Although minerals are the origin of most nutrients, and the bulk of most nutrient elements in the soil is held in crystalline form within primary and secondary minerals, they weather too slowly to support rapid plant growth. For example, The application of finely ground minerals, feldspar and apatite, to soil seldom provides the necessary amounts of potassium and phosphorus at a rate sufficient for good plant growth, as most of the nutrients remain bound in the crystals of those minerals.

The nutrients adsorbed onto the surfaces of clay colloids and soil organic matter provide a more accessible reservoir of many plant nutrients (e.g. K, Ca, Mg, P, Zn). As plants absorb the nutrients from the soil water, the soluble pool is replenished from the surface-bound pool. The decomposition of soil organic matter by microorganisms is another mechanism whereby the soluble pool of nutrients is replenished - this is important for the supply of plant-available N, S, P, and B from soil.

Gram for gram, the capacity of humus to hold nutrients and water is far greater than that of clay minerals. All in all, small amounts of humus may remarkably increase the soil's capacity to promote plant growth.

Soil Processes Important for Nutrient Uptake

Nutrients in the soil are taken up by the plant through its roots. To be taken up by a plant, a nutrient element must be located near the root surface; however, the supply of nutrients in contact with the root is rapidly depleted. There are three basic mechanisms whereby nutrient ions dissolved in the soil solution are brought into contact with plant roots:

1. Mass flow of water
2. Diffusion within water
3. Interception by root growth

All three mechanisms operate simultaneously, but one mechanism or another may be most important for a particular nutrient. For example, in the case of calcium, which is generally plentiful in the soil solution, mass flow alone can usually bring sufficient amounts to the root surface. However, in the case of phosphorus, diffusion is needed to supplement mass flow. For the most part, nutrient ions must travel some distance in the soil solution to reach the root surface. This movement can take place by mass flow, as when dissolved nutrients are carried along with the soil water flowing toward a root that is actively drawing water from the soil. In this type of movement, the nutrient ions are somewhat analogous to leaves floating down a stream. In addition, nutrient ions continually move by diffusion from areas of greater concentration toward the nutrient-depleted areas of lower concentration around the root surface. That process is due to random motion of molecules. By this means, plants can continue to take up nutrients even at night, when water is only slowly absorbed into the roots as transpiration has almost stopped. Finally, root interception comes into play as roots continually grow into new, undepleted soil.

Estimated relative importance of mass flow, diffusion and root interception as mechanisms in supplying plant nutrients to corn plant roots in soils

Nutrient	Approximate percentage supplied by:		
	Mass flow	Root interception	Diffusion
Nitrogen	98.8	1.2	0
Phosphorus	6.3	2.8	90.9
Potassium	20.0	2.3	77.7
Calcium	71.4	28.6	0
Sulfur	95.0	5.0	0
Molybdenum	95.2	4.8	0

In the above table, phosphorus and potassium nutrients move more by diffusion than they do by mass flow in the soil water solution, as they are rapidly taken up by the roots creating a concentration of almost zero near the roots (the plants cannot transpire enough water to draw more of those nutrients near the roots). The very steep concentration gradient is of greater influence in the movement of those ions than is the movement of those by mass flow. The movement by mass flow requires the transpiration of water from the plant causing water and solution ions to also move toward the roots. Movement by root interception is slowest as the plants must extend their roots.

Plants move ions out of their roots in an effort to move nutrients in from the soil. Hydrogen H^+ is exchanged for other cations, and carbonate (HCO_3^-) and hydroxide (OH^-) anions are exchanged for nutrient anions. As plant roots remove nutrients from the soil water solution, they are replenished as other ions move off of clay and humus (by ion exchange or desorption), are added from the weathering of soil minerals, and are released by the decomposition of soil organic matter. Plants derive a large proportion of their anion nutrients from decomposing organic matter, which typically holds about 95 percent of the soil nitrogen, 5 to 60 percent of the soil phosphorus and about 80 percent of the soil sulfur. Where crops are produced, the replenishment of nutrients in the soil must usually be augmented by the addition of fertilizer or organic matter.

Because nutrient uptake is an active metabolic process, conditions that inhibit root metabolism may also inhibit nutrient uptake. Examples of such conditions include waterlogging or soil compaction resulting in poor soil aeration, excessively high or low soil temperatures, and above-ground conditions that result in low translocation of sugars to plant roots.

Carbon

Measuring soil respiration in the field using an SRS2000 system.

Plants obtain their carbon from atmospheric carbon dioxide. About 45% of a plant's dry mass is carbon; plant residues typically have a carbon to nitrogen ratio (C/N) of between 13:1 and 100:1. As the soil organic material is digested by arthropods and micro-organisms, the C/N decreases as the carbonaceous material is metabolized and carbon dioxide (CO_2) is released as a byproduct which then finds its way out of the soil and into the atmosphere. The nitrogen is sequestered in the bodies of the living matter of those decomposing organisms and so it builds up in the soil. Normal CO_2 concentration in the atmosphere is 0.03%, this can be the factor limiting plant growth. In a field of maize on a still day during high light conditions in the growing season, the CO_2 concentration drops very low, but under such conditions the crop could use up to 20 times the normal concentration. The respiration of CO_2 by soil micro-organisms decomposing soil organic matter contributes an important amount of CO_2 to the photosynthesising plants. Within the soil, CO_2 concentration is 10 to 100 times that of atmospheric levels but may rise to toxic levels if the soil porosity is low or if diffusion is impeded by flooding.

Nitrogen

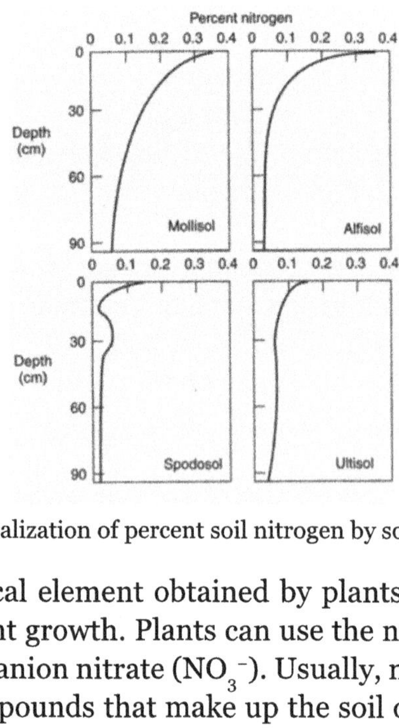

Generalization of percent soil nitrogen by soil order

Nitrogen is the most critical element obtained by plants from the soil and nitrogen deficiency often limits plant growth. Plants can use the nitrogen as either the ammonium cation (NH_4^+) or the anion nitrate (NO_3^-). Usually, most of the nitrogen in soil is bound within organic compounds that make up the soil organic matter, and must be mineralized to the ammonium or nitrate form before it can be taken up by most plants. The total nitrogen content depends largely on the soil organic matter content, which in turn depends on the climate, vegetation, topography, age and soil management. Soil nitrogen typically decreases by 0.2 to 0.3% for every temperature increase by 10 °C. Usually, grassland soils contain more soil nitrogen than forest soils. Cultivation decreases soil nitrogen by exposing soil organic matter to decomposition by microorganisms, and soils under no-tillage maintain more soil nitrogen than tilled soils.

Some micro-organisms are able to metabolise organic matter and release ammonium in a process called *mineralisation*. Others take free ammonium and oxidise it to nitrate. Nitrogen-fixing bacteria are capable of metabolising N_2 into the form of ammonia in a process called nitrogen fixation. Both ammonium and nitrate can be *immobilized* by their incorporation into the microbes' living cells, where it is temporarily sequestered in the form of amino acids and protein. Nitrate may also be lost from the soil when bacteria metabolise it to the gases N_2 and N_2O. The loss of gaseous forms of nitrogen to the atmosphere due to microbial action is called *denitrification*. Nitrogen may also be *leached* from the soil if it is in the form of nitrate or lost to the atmosphere as ammonia due to a chemical reaction of ammonium with alkaline soil by way of a process called *volatilisation*. Ammonium may also be sequestered in clay by *fixation*. A small amount of nitrogen is added to soil by rainfall.

Nitrogen Gains

In the process of mineralisation, microbes feed on organic matter, releasing ammonia (NH_3), ammonium (NH_4^+) and other nutrients. As long as the carbon to nitrogen ratio (C/N) of fresh residues in the soil is above 30:1, nitrogen will be in short supply and other bacteria will feed on the ammonium and incorporate its nitrogen into their cells in the immobilization process. In that form the nitrogen is said to be *immobilised*. Later, when such bacteria die, they too are *mineralised* and some of the nitrogen is released as ammonium and nitrate. If the C/N is less than 15, ammonia is freed to the soil, where it may be used by bacteria which oxidise it to nitrate (nitrification). Bacteria may on average add 25 pounds (11 kg) nitrogen per acre, and in an unfertilised field, this is the most important source of usable nitrogen. In a soil with 5% organic matter perhaps 2 to 5% of that is released to the soil by such decomposition. It occurs fastest in warm, moist, well aerated soil. The mineralisation of 3% of the organic material of a soil that is 4% organic matter overall, would release 120 pounds (54 kg) of nitrogen as ammonium per acre.

Carbon/Nitrogen Ratio of Various Organic Materials	
Organic Material	**C:N Ratio**
Alfalfa	13
Bacteria	4
Clover, green sweet	16
Clover, mature sweet	23
Fungi	9
Forest litter	30
Humus in warm cultivated soils	11
Legume-grass hay	25
Legumes (alfalfa or clover), mature	20
Manure, cow	18
Manure, horse	16–45
Manure, human	10
Oat straw	80

Straw, cornstalks	90
Sawdust	250

In nitrogen fixation, rhizobium bacteria convert N_2 to ammonia (NH_3). Rhizobia share a symbiotic relationship with host plants, since rhizobia supply the host with nitrogen and the host provides rhizobia with nutrients and a safe environment. It is estimated that such symbiotic bacteria in the root nodules of legumes add 45 to 250 pounds of nitrogen per acre per year, which may be sufficient for the crop. Other, free-living nitrogen-fixing bacteria and blue-green algae live independently in the soil and release nitrate when their dead bodies are converted by way of mineralisation.

Some amount of usable nitrogen is fixed by lightning as nitric oxide (NO) and nitrogen dioxide (NO_2^-). Nitrogen dioxide is soluble in water to form nitric acid (HNO_3) solution of H^+ and NO_3^-. Ammonia, NH_3, previously released from the soil or from combustion, may fall with precipitation as nitric acid at a rate of about five pounds nitrogen per acre per year.

Nitrogen Sequestration

When bacteria feed on soluble forms of nitrogen (ammonium and nitrate), they temporarily sequester that nitrogen in their bodies in a process called *immobilisation*. At a later time when those bacteria die, their nitrogen may be released as ammonium by the processes of mineralisation.

Protein material is easily broken down, but the rate of its decomposition is slowed by its attachment to the crystalline structure of clay and when trapped between the clay layers. The layers are small enough that bacteria cannot enter. Some organisms can exude extracellular enzymes that can act on the sequestered proteins. However, those enzymes too may be trapped on the clay crystals.

Ammonium fixation occurs when ammonium pushes potassium ions from between the layers of clay such as illite or montmorillonite. Only a small fraction of soil nitrogen is held this way.

Nitrogen Losses

Usable nitrogen may be lost from soils when it is in the form of nitrate, as it is easily leached. Further losses of nitrogen occur by denitrification, the process whereby soil bacteria convert nitrate (NO_3^-) to nitrogen gas, N_2 or N_2O. This occurs when poor soil aeration limits free oxygen, forcing bacteria to use the oxygen in nitrate for their respiratory process. Denitrification increases when oxidisable organic material is available and when soils are warm and slightly acidic. Denitrification may vary throughout a soil as the aeration varies from place to place. Denitrification may cause the loss of 10 to 20 percent of the available nitrates within a day and when conditions are favourable to that process, losses of up to 60 percent of nitrate applied as fertiliser may occur.

Ammonium volatilisation occurs when ammonium reacts chemically with an alkaline soil, converting NH_4^+ to NH_3. The application of ammonium fertiliser to such a field can result in volatilisation losses of as much as 30 percent.

Phosphorus

After nitrogen, phosphorus is probably the element most likely to be deficient in soils. The soil mineral apatite is the most common mineral source of phosphorus. While there is on average 1000 lb of phosphorus per acre in the soil, it is generally unavailable in the form of phosphates of low solubility. Total phosphorus is about 0.1 percent by weight of the soil, but only one percent of that is available. Of the part available, more than half comes from the mineralisation of organic matter. Agricultural fields may need to be fertilised to make up for the phosphorus that has been removed in the crop.

When phosphorus does form solubilised ions of $H_2PO_4^-$, they rapidly form insoluble phosphates of calcium or hydrous oxides of iron and aluminum. Phosphorus is largely immobile in the soil and is not leached but actually builds up in the surface layer if not cropped. The application of soluble fertilisers to soils may result in zinc deficiencies as zinc phosphates form. Conversely, the application of zinc to soils may immobilise phosphorus again as zinc phosphate. Lack of phosphorus may interfere with the normal opening of the plant leaf stomata, resulting in plant temperatures 10 percent higher than normal. Phosphorus is most available when soil pH is 6.5 in mineral soils and 5.5 in organic soils.

Potassium

The amount of potassium in a soil may be as much as 80,000 lb per acre-foot, of which only 150 lb is available for plant growth. Common mineral sources of potassium are the mica biotite and potassium feldspar, $KAlSi_3O_8$. When solubilised, half will be held as exchangeable cations on clay while the other half is in the soil water solution. Potassium fixation often occurs when soils dry and the potassium is bonded between layers of illite clay. Under certain conditions, dependent on the soil texture, intensity of drying, and initial amount of exchangeable potassium, the fixed percentage may be as much as 90 percent within ten minutes. Potassium may be leached from soils low in clay.

Calcium

Calcium is one percent by weight of soils and is generally available but may be low as it is soluble and can be leached. It is thus low in sandy and heavily leached soil or strongly acidic mineral soil. Calcium is supplied to the plant in the form of exchangeable ions and moderately soluble minerals. Calcium is more available on the soil colloids than is potassium because the common mineral calcite, $CaCO_3$, is more soluble than potassium-bearing minerals.

Magnesium

Magnesium is one of the dominant exchangeable cations in most soils (as are calcium and potassium). Primary minerals that weather to release magnesium include hornblende, biotite and vermiculite. Soil magnesium concentrations are generally sufficient for optimal plant growth, but highly weathered and sandy soils may be magnesium deficient due to leaching by heavy precipitation.

Sulfur

Most sulfur is made available to plants, like phosphorus, by its release from decomposing organic matter. Deficiencies may exist in some soils (especially sandy soils) and if cropped, sulfur needs to be added. The application of large quantities of nitrogen to fields that have marginal amounts of sulfur may cause sulfur deficiency in the rapidly growing plants by the plant's growth outpacing the supply of sulfur. A 15-ton crop of onions uses up to 19 lb of sulfur and 4 tons of alfalfa uses 15 lb per acre. Sulfur abundance varies with depth. In a sample of soils in Ohio, United States, the sulfur abundance varied with depths, 0-6 inches, 6-12 inches, 12-18 inches, 18-24 inches in the amounts: 1056, 830, 686, 528 lb per acre respectively.

Micronutrients

The micronutrients essential for plant life, in their order of importance, include iron, manganese, zinc, copper, boron, chlorine and molybdenum. The term refers to plants' needs, not to their abundance in soil. They are required in very small amounts but are essential to plant health in that most are required parts of some enzyme system which speeds up plants' metabolisms. They are generally available in the mineral component of the soil, but the heavy application of phosphates can cause a deficiency in zinc and iron by the formation of insoluble zinc and iron phosphates. Iron deficiency may also result from excessive amounts of heavy metals or calcium minerals (lime) in the soil. Excess amounts of soluble boron, molybdenum and chloride are toxic.

Non-Essential Nutrients

Nutrients which enhance the health but whose deficiency does not stop the life cycle of plants include: cobalt, strontium, vanadium, silicon and nickel. As their importance are evaluated they may be added to the list of essential plant nutrients.

Soil Organic Matter

Soil organic matter is made up of organic compounds and includes plant, animal and microbial material, both living and dead. A typical soil has a biomass composition of 70% microorganisms, 22% macrofauna, and 8% roots. The living component of an acre of soil may include 900 lb of earthworms, 2400 lb of fungi, 1500 lb of bacteria, 133 lb of protozoa and 890 lb of arthropods and algae.

A small part of the organic matter consists of the living cells such as bacteria, molds, and actinomycetes that work to break down the dead organic matter. Were it not for the action of these micro-organisms, the entire carbon dioxide part of the atmosphere would be sequestered as organic matter in the soil.

Chemically, organic matter is classed as follows:

1. Polysaccharides
    1. cellulose
    2. hemicellulose
    3. starch
    4. pectin
2. Lignins
3. Proteins

Most living things in soils, including plants, insects, bacteria, and fungi, are dependent on organic matter for nutrients and/or energy. Soils have organic compounds in varying degrees of decomposition which rate is dependent on the temperature, soil moisture, and aeration. Bacteria and fungi feed on the raw organic matter, which are fed upon by amoebas, which in turn are fed upon by nematodes and arthropods. Organic matter holds soils open, allowing the infiltration of air and water, and may hold as much as twice its weight in water. Many soils, including desert and rocky-gravel soils, have little or no organic matter. Soils that are all organic matter, such as peat (histosols), are infertile. In its earliest stage of decomposition, the original organic material is often called raw organic matter. The final stage of decomposition is called humus.

In grassland, much of the organic matter added to the soil is from the deep, fibrous, grass root systems. By contrast, tree leaves falling on the forest floor are the principal source of soil organic matter in the forest. Another difference is the frequent occurrence in the grasslands of fires that destroy large amounts of aboveground material but stimulate even greater contributions from roots. Also, the much greater acidity under any forests inhibits the action of certain soil organisms that otherwise would mix much of the surface litter into the mineral soil. As a result, the soils under grasslands generally develop a thicker A horizon with a deeper distribution of organic matter than in comparable soils under forests, which characteristically store most of their organic matter in the forest floor (O horizon) and thin A horizon.

Humus

Humus refers to organic matter that has been decomposed by soil flora and fauna to the point where it is resistant to further breakdown. Humus usually constitutes only five percent of the soil or less by volume, but it is an essential source of nutrients and adds important textural qualities crucial to soil health and plant growth. Humus also hold bits of undecomposed organic matter which feed arthropods and worms which further improve the soil. The end product, humus, is soluble in water and forms a weak acid that can attack silicate minerals. Humus is a colloid with a high cation and anion exchange capacity that on a dry weight basis is many times greater than that of clay colloids. It also acts as a buffer, like clay, against changes in pH and soil moisture.

Humic acids and fulvic acids, which begin as raw organic matter, are important constituents of humus. After the death of plants and animals, microbes begin to feed on the residues, resulting finally in the formation of humus. With decomposition, there is a reduction of water-soluble constituents, cellulose and hemicellulose, and nutrients such as nitrogen, phosphorus, and sulfur. As the residues break down, only stable molecules made of aromatic carbon rings, oxygen and hydrogen remain in the form of humin, lignin and lignin complexes collectively called humus. While the structure of humus has few nutrients, it is able to attract and hold cation and anion nutrients by weak bonds that can be released into the soil solution in response to changes in soil pH.

Lignin is resistant to breakdown and accumulates within the soil. It also reacts with amino acids, which further increases its resistance to decomposition, including enzymatic decomposition by microbes. Fats and waxes from plant matter have some resistance to decomposition and persist in soils for a while. Clay soils often have higher organic contents that persist longer than soils without clay as the organic molecules adhere to and are stabilised by the clay. Proteins normally decompose readily, but when bound to clay particles, they become more resistant to decomposition. Clay particles also absorb the enzymes exuded by microbes which would normally break down proteins. The addition of organic matter to clay soils can render that organic matter and any added nutrients inaccessible to plants and microbes for many years. High soil tannin (polyphenol) content can cause nitrogen to be sequestered in proteins or cause nitrogen immobilisation.

Humus formation is a process dependent on the amount of plant material added each year and the type of base soil. Both are affected by climate and the type of organisms present. Soils with humus can vary in nitrogen content but typically have 3 to 6 percent nitrogen. Raw organic matter, as a reserve of nitrogen and phosphorus, is a vital component affecting soil fertility. Humus also absorbs water, and expands and shrinks between dry and wet states, increasing soil porosity. Humus is less stable than the soil's mineral constituents, as it is reduced by microbial decomposition, and over time its concentration diminshes without the addition of new organic matter. However, humus may persist over centuries if not millennia.

Climate and Organic Matter

The production, accumulation and degradation of organic matter are greatly dependent on climate. Temperature, soil moisture and topography are the major factors affecting the accumulation of organic matter in soils. Organic matter tends to accumulate under wet or cold conditions where decomposer activity is impeded by low temperature or excess moisture which results in anaerobic conditions. Conversely, excessive rain and high temperatures of tropical climates enables rapid decomposition of organic matter and leaching of plant nutrients; forest ecosystems on these soils rely on efficient recycling of nutrients and plant matter to maintain their productivity. Excessive slope may encourage the erosion of the top layer of soil which holds most of the raw organic material that would otherwise eventually become humus.

Plant Residue in Soil

Cellulose and hemicellulose undergo fast decomposition by fungi and bacteria, with a half-life of 12–18 days in a temperate climate. Brown rot fungi can decompose the cellulose and hemicellulose, leaving the lignin and phenolic compounds behind. Starch, which is an energy storage system for plants, undergoes fast decomposition by bacteria and fungi. Lignin consists of polymers composed of 500 to 600 units with a highly branched, amorphous structure. Lignin undergoes very slow decomposition, mainly by white rot fungi and actinomycetes; its half-life under temperate conditions is about six months.

Soil Horizons

A horizontal layer of the soil, whose physical features, composition and age are distinct from those above and beneath, are referred to as a soil horizon. The naming of a horizon is based on the type of material of which it is composed. Those materials reflect the duration of specific processes of soil formation. They are labelled using a shorthand notation of letters and numbers which describe the horizon in terms of its colour, size, texture, structure, consistency, root quantity, pH, voids, boundary characteristics and presence of nodules or concretions. Few soil profiles have all the major horizons. Some may have only one horizon.

The exposure of parent material to favourable conditions produces mineral soils that are marginally suitable for plant growth. That growth often results in the accumulation of organic residues. The accumulated organic layer called the O horizon produces a more active soil due to the effect of the organisms that live within it. Organisms colonise and break down organic materials, making available nutrients upon which other plants and animals can live. After sufficient time, humus moves downward and is deposited in a distinctive organic surface layer called the A horizon.

Classification

Soil is classified into categories in order to understand relationships between different soils and to determine the suitability of a soil for a particular use. One of the first classification systems was developed by the Russian scientist Dokuchaev around 1880. It was modified a number of times by American and European researchers, and developed into the system commonly used until the 1960s. It was based on the idea that soils have a particular morphology based on the materials and factors that form them. In the 1960s, a different classification system began to emerge which focused on soil morphology instead of parental materials and soil-forming factors. Since then it has undergone further modifications. The World Reference Base for Soil Resources (WRB) aims to establish an international reference base for soil classification.

Soil Classification Systems

Australia

There are fourteen soil orders at the top level of the Australian Soil Classification. They are: Anthroposols, Organosols, Podosols, Vertosols, Hydrosols, Kurosols, Sodosols, Chromosols, Calcarosols, Ferrosols, Dermosols, Kandosols, Rudosols and Tenosols.

USA

A taxonomy is an arrangement in a systematic manner; the USDA soil taxonomy has six levels of classification. They are, from most general to specific: order, suborder, great group, subgroup, family and series. Soil properties that can be measured quantitatively are used in this classification system — they include: depth, moisture, temperature, texture, structure, cation exchange capacity, base saturation, clay mineralogy, organic matter content and salt content. There are 12 soil orders (the top hierarchical level) in the USDA soil taxonomy. The names of the orders end with the suffix *-sol*. The criteria for the different soil orders include properties that reflect major differences in the genesis of soils. The orders are:

- Alfisol - soils with aluminium and iron. They have horizons of clay accumulation, and form where there is enough moisture and warmth for at least three months of plant growth. They constitute 10.1% of soils worldwide.

- Andisol - volcanic ash soils. They are young and very fertile. They cover 1% of the world's ice-free surface.

- Aridisol - dry soils forming under desert conditions which have fewer than 90 consecutive days of moisture during the growing season and are nonleached. They include nearly 12% of soils on Earth. Soil formation is slow, and accumulated organic matter is scarce. They may have subsurface zones of caliche or duripan. Many aridisols have well-developed Bt horizons showing clay movement from past periods of greater moisture.

- Entisol - recently formed soils that lack well-developed horizons. Commonly found on unconsolidated river and beach sediments of sand and clay or volcanic ash, some have an A horizon on top of bedrock. They are 18% of soils worldwide.
- Gelisol - permafrost soils with permafrost within two metres of the surface or gelic materials and permafrost within one metre. They constitute 9.1% of soils worldwide.
- Histosol - organic soils, formerly called bog soils, are 1.2% of soils worldwide.
- Inceptisol - young soils. They have subsurface horizon formation but show little eluviation and illuviation. They constitute 15% of soils worldwide.
- Mollisol - soft, deep, dark fertile soil formed in grasslands and some hardwood forests with very thick A horizons. They are 7% of soils worldwide.
- Oxisol - are heavily weathered, are rich in iron and aluminum oxides (sesquioxides) or kaolin but low in silica. They have only trace nutrients due to heavy tropical rainfall and high temperatures and low CEC of the remaining clays. They are 7.5% of soils worldwide.
- Spodosol - acid soils with organic colloid layer complexed with iron and aluminium leached from a layer above. They are typical soils of coniferous and deciduous forests in cooler climates. They constitute 4% of soils worldwide.
- Ultisol - acid soils in the humid tropics and subtropics, which are depleted in calcium, magnesium and potassium (important plant nutrients). They are highly weathered, but not as weathered as Oxisols. They make up 8% of the soil worldwide.
- Vertisol - inverted soils. They are clay-rich and tend to swell when wet and shrink upon drying, often forming deep cracks into which surface layers can fall. They are difficult to farm or to construct roads and buildings due to their high expansion rate. They constitute 2.4% of soils worldwide.

Alfisol

Vertisol

The percentages listed above are for land area free of ice. “Soils of Mountains”, which constitute the balance (11.6%), have a mixture of those listed above, or are classified as “Rugged Mountains” which have no soil.

The above soil orders in sequence of increasing degree of development are Entisols, Inceptisols, Aridisols, Mollisols, Alfisols, Spodosols, Ultisols, and Oxisols. Histosols and Vertisols may appear in any of the above at any time during their development.

The soil suborders within an order are differentiated on the basis of soil properties and horizons which depend on soil moisture and temperature. Forty-seven suborders are recognized in the United States.

The soil great group category is a subdivision of a suborder in which the kind and sequence of soil horizons distinguish one soil from another. About 185 great groups are recognized in the United States. Horizons marked by clay, iron, humus and hard pans and soil features such as the expansion-contraction of clays (that produce self-mixing provided by clay), temperature, and marked quantities of various salts are used as distinguishing features.

The great group categories are divided into three kinds of soil subgroups: typic, intergrade and extragrade. A typic subgroup represents the basic or ‘typical’ concept of the great group to which the described subgroup belongs. An intergrade subgroup describes the properties that suggest how it grades towards (is similar to) soils of other soil great groups, suborders or orders. These properties are not developed or expressed well enough to cause the soil to be included within the great group towards which they grade, but suggest similarities. Extragrade features are aberrant properties which prevent that soil from being included in another soil classification. About 1,000 soil subgroups are defined in the United States.

A soil family category is a group of soils within a subgroup and describes the physical and chemical properties which affect the response of soil to agricultural management and engineering applications. The principal characteristics used to differentiate soil families include texture, mineralogy, pH, permeability, structure, consistency, the locale's precipitation pattern, and soil temperature. For some soils the criteria also specify the percentage of silt, sand and coarse fragments such as gravel, cobbles and rocks. About 4,500 soil families are recognised in the United States.

A family may contain several soil series which describe the physical location using the name of a prominent physical feature such as a river or town near where the soil sample was taken. An example would be Merrimac for the Merrimack River in New Hampshire, USA. More than 14,000 soil series are recognised in the United States. This permits very specific descriptions of soils.

A soil phase of series, originally called 'soil type' describes the soil surface texture, slope, stoniness, saltiness, erosion, and other conditions.

Uses

Soil is used in agriculture, where it serves as the anchor and primary nutrient base for plants; however, as demonstrated by hydroponics, it is not essential to plant growth if the soil-contained nutrients can be dissolved in a solution. The types of soil and available moisture determine the species of plants that can be cultivated.

Soil material is also a critical component in the mining, construction and landscape development industries. Soil serves as a foundation for most construction projects. The movement of massive volumes of soil can be involved in surface mining, road building and dam construction. Earth sheltering is the architectural practice of using soil for external thermal mass against building walls. Many building materials are soil based.

Soil resources are critical to the environment, as well as to food and fibre production. Soil provides minerals and water to plants. Soil absorbs rainwater and releases it later, thus preventing floods and drought. Soil cleans water as it percolates through it. Soil is the habitat for many organisms: the major part of known and unknown biodiversity is in the soil, in the form of invertebrates (earthworms, woodlice, millipedes, centipedes, snails, slugs, mites, springtails, enchytraeids, nematodes, protists), bacteria, archaea, fungi and algae; and most organisms living above ground have part of them (plants) or spend part of their life cycle (insects) below-ground. Above-ground and below-ground biodiversities are tightly interconnected, making soil protection of paramount importance for any restoration or conservation plan.

The biological component of soil is an extremely important carbon sink since about 57% of the biotic content is carbon. Even on desert crusts, cyanobacteria, lichens and mosses capture and sequester a significant amount of carbon by photosynthesis. Poor farming and grazing methods have degraded soils and released much of this sequestered carbon to the atmosphere. Restoring the world's soils could offset some of the huge increase in greenhouse gases causing global warming, while improving crop yields and reducing water needs.

Waste management often has a soil component. Septic drain fields treat septic tank effluent using aerobic soil processes. Landfills use soil for daily cover. Land application of waste water relies on soil biology to aerobically treat BOD.

Organic soils, especially peat, serve as a significant fuel resource; but wide areas of peat production, such as sphagnum bogs, are now protected because of patrimonial interest.

Geophagy is the practice of eating soil-like substances. Both animals and human cultures occasionally consume soil for medicinal, recreational, or religious purposes. It has been shown that some monkeys consume soil, together with their preferred food (tree foliage and fruits), in order to alleviate tannin toxicity.

Soils filter and purify water and affect its chemistry. Rain water and pooled water from ponds, lakes and rivers percolate through the soil horizons and the upper rock strata, thus becoming groundwater. Pests (viruses) and pollutants, such as persistent organic pollutants (chlorinated pesticides, polychlorinated biphenyls), oils (hydrocarbons), heavy metals (lead, zinc, cadmium), and excess nutrients (nitrates, sulfates, phosphates) are filtered out by the soil. Soil organisms metabolise them or immobilise them in their biomass and necromass, thereby incorporating them into stable humus. The physical integrity of soil is also a prerequisite for avoiding landslides in rugged landscapes.

Degradation

Land degradation refers to a human-induced or natural process which impairs the capacity of land to function. Soils degradation involves the acidification, contamination, desertification, erosion or salination.

Soil acidification is beneficial in the case of alkaline soils, but it degrades land when it lowers crop productivity and increases soil vulnerability to contamination and erosion. Soils are often initially acid because their parent materials were acid and initially low in the basic cations (calcium, magnesium, potassium and sodium). Acidification occurs when these elements are leached from the soil profile by rainfall or by the harvesting of forest or agricultural crops. Soil acidification is accelerated by the use of acid-forming nitrogenous fertilizers and by the effects of acid precipitation.

Soil contamination at low levels is often within a soil's capacity to treat and assimilate waste material. Soil biota can treat waste by transforming it; soil colloids can adsorb the waste material. Many waste treatment processes rely on this treatment capacity. Exceeding treatment capacity can damage soil biota and limit soil function. Derelict soils occur where industrial contamination or other development activity damages the soil to such a degree that the land cannot be used safely or productively. Remediation of derelict soil uses principles of geology, physics, chemistry and biology to degrade, attenuate, isolate or remove soil contaminants to restore soil functions and values. Techniques include leaching, air sparging, chemical amendments, phytoremediation, bioremediation and natural degradation.

Desertification is an environmental process of ecosystem degradation in arid and semi-arid regions, often caused by human activity. It is a common misconception that droughts cause desertification. Droughts are common in arid and semiarid lands. Well-managed lands can recover from drought when the rains return. Soil management tools include maintaining soil nutrient and organic matter levels, reduced tillage and increased cover. These practices help to control erosion and maintain productivity during periods when moisture is available. Continued land abuse during droughts, however, increases land degradation. Increased population and livestock pressure on marginal lands accelerates desertification.

Erosion of soil is caused by water, wind, ice, and movement in response to gravity. More than one kind of erosion can occur simultaneously. Erosion is distinguished from weathering, since erosion also transports eroded soil away from its place of origin (soil in transit may be described as sediment). Erosion is an intrinsic natural process, but in many places it is greatly increased by human activity, especially poor land use practices. These include agricultural activities which leave the soil bare during times of heavy rain or strong winds, overgrazing, deforestation, and improper construction activity. Improved management can limit erosion. Soil conservation techniques which are employed include changes of land use (such as replacing erosion-prone crops with grass or other soil-binding plants), changes to the timing or type of agricultural operations, terrace building, use of erosion-suppressing cover materials (including cover crops and other plants), limiting disturbance during construction, and avoiding construction during erosion-prone periods.

Desertification

A serious and long-running water erosion problem occurs in China, on the middle reaches of the Yellow River and the upper reaches of the Yangtze River. From the Yellow River, over 1.6 billion tons of sediment flow each year into the ocean. The sediment originates primarily from water erosion (gully erosion) in the Loess Plateau region of northwest China.

Soil piping is a particular form of soil erosion that occurs below the soil surface. It causes levee and dam failure, as well as sink hole formation. Turbulent flow removes soil starting at the mouth of the seep flow and the subsoil erosion advances up-gradient. The term sand boil is used to describe the appearance of the discharging end of an active soil pipe.

Soil salination is the accumulation of free salts to such an extent that it leads to degradation of the agricultural value of soils and vegetation. Consequences include corrosion damage, reduced plant growth, erosion due to loss of plant cover and soil structure, and water quality problems due to sedimentation. Salination occurs due to a combination of natural and human-caused processes. Arid conditions favour salt accumulation. This is especially apparent when soil parent material is saline. Irrigation of arid lands is especially problematic. All irrigation water has some level of salinity. Irrigation, especially when it involves leakage from canals and overirrigation in the field, often raises the underlying water table. Rapid salination occurs when the land surface is within the capillary fringe of saline groundwater. Soil salinity control involves watertable control and flushing with higher levels of applied water in combination with tile drainage or another form of subsurface drainage.

Erosion control

Reclamation

Soils which contain high levels of particular clays, such as smectites, are often very fertile. For example, the smectite-rich clays of Thailand's Central Plains are among the most productive in the world.

Many farmers in tropical areas, however, struggle to retain organic matter in the soils they work. In recent years, for example, productivity has declined in the low-clay soils of northern Thailand. Farmers initially responded by adding organic matter from termite mounds, but this was unsustainable in the long-term. Scientists experimented with adding bentonite, one of the smectite family of clays, to the soil. In field trials, conducted by scientists from the International Water Management Institute in cooperation with Khon Kaen University and local farmers, this had the effect of helping retain water and nutrients. Supplementing the farmer's usual practice with a single application of 200 kg bentonite per rai (6.26 rai = 1 hectare) resulted in an average yield increase of 73%. More work showed that applying bentonite to degraded sandy soils reduced the risk of crop failure during drought years.

In 2008, three years after the initial trials, IWMI scientists conducted a survey among 250 farmers in northeast Thailand, half of whom had applied bentonite to their fields. The average improvement for those using the clay addition was 18% higher than for non-clay users. Using the clay had enabled some farmers to switch to growing vegetables, which need more fertile soil. This helped to increase their income. The researchers estimated that 200 farmers in northeast Thailand and 400 in Cambodia had adopted the use of clays, and that a further 20,000 farmers were introduced to the new technique.

If the soil is too high in clay, adding gypsum, washed river sand and organic matter will balance the composition. Adding organic matter (like ramial chipped wood for instance) to soil which is depleted in nutrients and too high in sand will boost its quality.

Soil Structure

Soil structure describes the arrangement of the solid parts of the soil and of the pore space located between them. It is determined by how individual soil granules clump or bind together and aggregate, and therefore, the arrangement of soil pores between them. Soil structure has a major influence on water and air movement, biological activity, root growth and seedling emergence.

Overview

Soil structure describes the arrangement of the solid parts of the soil and of the pore space located between them (Marshall & Holmes, 1979). The structure depends on what the soil developed from. The quality of soil structure will decline under most forms of cultivation—the associated mechanical mixing of the soil compacts and shears aggregates and fills pore spaces; it also exposes organic matter to a greater rate of decay and oxidation (Young & Young, 2001). A further consequence of continued cultivation and traffic is the development of compacted, impermeable layers or pans within the profile.

The decline of soil structure under irrigation is usually related to the breakdown of aggregates and dispersion of clay material as a result of rapid wetting. This is particularly so if soils are sodic; that is, having a high exchangeable sodium percentage (ESP) of the cations attached to the clays. High sodium levels (compared to high calcium levels) cause particles to repel one another when wet, and the associated aggregates to disaggregate and disperse. The ESP will increase if irrigation causes salty water (even of low concentration) to gain access to the soil.

A wide range of practices are undertaken to preserve and improve soil structure. For example, the NSW Department of Land and Water Conservation, (1991) advocates: increasing organic content by incorporating pasture phases into cropping rotations; reducing or eliminating tillage and cultivation in cropping and pasture activities; avoiding soil disturbance during periods of excessive dry or wet when soils may accordingly tend to shatter or smear; and ensuring sufficient ground cover to protect the soil from raindrop impact. In irrigated agriculture, it may be recommended to: apply gypsum (calcium sulfate) to displace sodium cations with calcium and so reduce ESP or sodicity, avoid rapid wetting, and avoid disturbing soils when too wet or dry.

Impacts of Improving Soil Structure

The benefits of improving soil structure for the growth of plants, particularly in an agricultural setting include: reduced erosion due to greater soil aggregate strength and decreased overland flow; improved root penetration and access to soil moisture and nutrients; improved emergence of seedlings due to reduced crusting of the surface; and greater water infiltration, retention and availability due to improved porosity.

It has been estimated that productivity from irrigated perennial horticulture could be increased by two to three times the present level by improving soil structure, because of the resulting access by plants to available soil water and nutrients (Cockroft & Olsson, 2000, cited in Land and Water Australia 2007). The NSW Department of Land and Water Conservation (1991) suggests that in cropping systems, wheat yields can be increased by 10 kg/ha for every extra millimetre of rain that is able to infiltrate due to soil structure improvements.

Hardsetting Soil

Hardsetting soils lose their structure when wet and then set hard as they dry out to form a structureless mass that is very difficult to cultivate. They can only be tilled when their moisture content is within a limited range. When they are tilled the result is often a very cloddy surface (poor tilth). As they dry out the high soil strength often restricts seedling and root growth. Infiltration rates are low and runoff of rain and irrigation limits the productivity of many hardsetting soils.

Definition

Hardsetting has been defined this way: "*A hardsetting soil is one that sets to an almost homogeneous mass on drying. It may have occasional cracks, typically at a spacing of >0.1 m. Air dry hardset soil is hard and brittle, and it is not possible to push a forefinger into the profile face. Typically, it has a tensile strength of 90 kN* m^{-2}. *Soils that crust are not necessarily hardsetting since a hardsetting horizon is thicker than a crust. (In cultivated soils the thickness of the hardsetting horizon is frequently equal to or greater than that of the cultivated layer.) Hardsetting soil is not permanently cemented and is soft when wet. The clods in a hardsetting horizon that has been cultivated will partially or totally disintegrate upon wetting. If the soil has been sufficiently wetted, it will revert to its hardset state on drying. This can happen after flood irrigation or a single intense rainfall event.*"

Topsoil

Surface runoff, a type of nonpoint source pollution, from a farm field in Iowa during a rain storm. Topsoil as well as farm fertilizers and other potential pollutants run off unprotected farm fields when heavy rains occur.

Topsoil is the upper, outermost layer of soil, usually the top 2 inches (5.1 cm) to 8 inches (20 cm). It has the highest concentration of organic matter and microorganisms and is where most of the Earth's biological soil activity occurs. Four elements constitute the composition of soil. Those elements are mineral particles, organic matter, water, and air. The majority of the top soils' volume consists of 50 to 80 percent of these particles which form the skeletal structure of most soils. This composition allows the soil to sustain its own weight, and other internal matter such as water and overlying landscape. Organic matter, another important element, varies on quantity on different soils. This provokes positive and negative effects or reactions on the soil. The strength of soil structure decreases with the presence of organic matter, creating weak bearing capacities. Organic matter condenses and settles in different ways under certain conditions, such as roadbeds and foundations. The skeletal structure becomes affected once the soil is dewatered. The soil's volume substantially decreases. It decomposes and suffers wind erosion.

Importance

Plants generally concentrate their roots in and obtain most of their vital nutrients from this layer. Actual depth of the topsoil layer can be measured as the depth from the surface to the first densely packed soil layer known as subsoil.

Classification

In soil classification systems, topsoil is known as the "O Horizon or A Horizon," therefore, it is the very top layer.

Commercially available topsoil (manufactured or naturally occurring) in the United Kingdom should be classified to British Standard BS 3882 with the current version dated 2015. The standard has several classifications of topsoil with the final classification requiring material to meet certain threshold criteria such as Nutrient Content, Extractable Phytotoxic Elements, Particle Size Distribution, Organic Matter Content, Carbon:Nitrogen ratio, Electrical Conductivity, Loss on Ignition, pH, Chemical and Physical Contamination. The topsoil should be sampled in accordance with the British Standard and European Norm BS EN 12579:2013 Soil improvers and growing media - Sampling. During construction of garden areas for housing plots the topsoil should be underlain by a layer of suitably certified subsoil that conforms to the British Standard BS 8601:2013 Specification for subsoil and requirements for use.

It is always recommended that for construction projects that topsoil is placed in accordance with the DEFRA report Construction Code of Practice for the Sustainable Use of Soils on Construction Sites

Evaluation

When starting a gardening project, it is crucial to check whether or not the soil is satisfactory. Different types of plants vary in their nutrient needs and preferred soil conditions, many are strongly adapted to particular conditions. However, some general guidelines for "desired levels of Topsoil nutrients" have been made, broadly suitable for many plants.

Category	Desired Results
pH Level	5.0 to 6.2
Phosphorus (P-I)	Index of 50
Potassium (K-I)	Index of 50
Calcium (Ca%)	40-60% of Cation Exchange Capacity (CEC)
Magnesium (Mg%)	8-10% of CEC
Base saturation (BS%)	35-80% of CEC
Manganese (Mn-I)	Index > 25
Zinc (Zn-I)	Index > 25
Copper (Cu-I)	Index > 25

The two common types of Topsoil are Bulk and Bagged Topsoil. The following table illustrates major differences between the two.

Topsoil Type	HM%	BS%	pH	P-I	K-I	Ca%	Mg%
Bulk	0.3	69	5.2	009	026	45	10
Bagged	0.7	78	5.8	166+	178	56	12.3

Alternatively the British Standard relates to other working values:

Category	Desired Results
pH Level	5.5 to 8.5
Phosphate (PO4)	16 to 140 mg/L
Potassium (K)	121 to 1500 mg/L
Magnesium (Mg)	51 to 600 mg/L
Nickel (Ni)	from <60 mg/kg
Zinc (Zn)	from <200 mg/kg
Copper (Cu)	from <100 mg/kg

This is for a multipurpose grade and certain levels can alter with regard to soil pH. Other uses specified in the standard that allows for a variety of uses in different and specific scenarios includes:

Acidic, Calcareous, Low Fertility, Low Fertility Acidic and Low Fertility Calcareous. These uses are limited to specific site scenarios and acceptance should be on a case by case basis for construction projects.

The Carbon Nitrogen Ratio

Topsoil is the primary resource for plants to grow and crops to thrive and the main two parameters for this are Carbon and Nitrogen. The Carbon provides energy and

Nitrogen is a tissue builder and plants require them in a range of ratios to enable suitable growth. An optimum figure for Topsoil in the UK is a ratio of less than 20:1. This ensures that the soil has a suitable energy reserve as well as tissue building material to enable the plants to thrive. A sawdust typically has a carbonaceous base and this a high C:N ratio (in the order of c. 400:1) while an Alfalfa Hay has a low carbonaceous content and can typically have a C:N ratio in the order of 12:1.

Commercial Application

A variety of soil mixtures are sold commercially as topsoil, usually for use in improving gardens and lawns, e.g. container gardens, potting soil and peat. Another important yet not commonly known use for topsoil is for proper surface grading near residential buildings such as homes. "The ground around the home should slope down six inches for the first ten feet away from the home. This can often be done by adding topsoil (not sand or gravel)."

Erosion

A major environmental concern known as topsoil erosion occurs when the topsoil layer is blown or washed away. Without topsoil, little plant life is possible. The estimated annual costs of public and environmental health losses related to soil erosion exceed $45 billion. Conventional agriculture encourages the depletion of topsoil because the soil must be plowed and replanted each year. Sustainable techniques attempt to slow erosion through the use of cover crops in order to build organic matter in the soil. The United States alone loses almost 3 tons of topsoil per acre per year. This is of great ecological concern as one inch of topsoil can take between 500 and 1,000 years to form naturally. On current trends, the world has about 60 years of topsoil left.

Subsoil

Subsoil layer

Subsoil is the layer of soil under the topsoil on the surface of the ground. Like topsoil it is composed of a variable mixture of small particles such as sand, silt and/or clay, but it lacks the organic matter and humus content of topsoil. Below the subsoil is the substratum, which can be residual bedrock, sediments, or aeolian deposits. As it is lacking in dark humus, subsoil is usually paler in colour than the overlying topsoil. It may contain the deeper roots of some plants, such as trees, but a majority of plant roots lie within the surface topsoil.

Sources

- *Håkansson, Inge; Reeder, Randall C. (March 1994). "Subsoil compaction by vehicles with high axle load—extent, persistence and crop response". Volume 29, Issues 2–3. Soil and Tillage Research. pp. 277–304. Retrieved October 11, 2012.* (subscription required)
- *Adams, Fred; . Moore, B. L (January 1983). "Chemical Factors Affecting Root Growth in Subsoil Horizons of Coastal Plain Soils". Vol. 47 No. 1. Soil Science Society of America Journal. pp. 99–102. Retrieved October 11, 2012.* (subscription required)

Paleosol

Paleosols sequence, Tuscany, Italy

In the geosciences, paleosol (*palaeosol* in Great Britain and Australia) can have two meanings. The first meaning, common in geology and paleontology, refers to a former soil preserved by burial underneath either sediments (alluvium or loess) or volcanic deposits (volcanic ash), which in the case of older deposits have lithified into rock. In Quaternary geology, sedimentology, paleoclimatology, and geology in general, it is the typical and accepted practice to use the term "paleosol" to designate such "fossil" soils found buried within either sedimentary or volcanic deposits exposed in all continents as illustrated by Rettallack (2001), Kraus (1999), and innumerable other published papers and books.

Etched section of paleosol from the Gulf, San Salvador Island, Bahamas, indicating the top of the Pleistocene Grotto Beach Formation (limestone).

In soil science, paleosols are soils formed long periods ago that have no relationship in their chemical and physical characteristics to the present-day climate or vegetation. Such soils form on extremely old continental cratons and as small scattered localities in outliers of ancient rock. Because of the changes in the Earth's climate over the last fifty million years, soils formed under tropical rainforest (or even savanna) have become exposed to increasingly arid climates which cause former oxisols, ultisols or even alfisols to dry out in such a manner that a very hard crust is formed. This process has occurred so extensively in most parts of Australia as to restrict soil development - the former soil is effectively the parent material for a new soil, but it is so unweatherable that only a very poorly developed soil can exist in present dry climates, especially when they have become much drier during glacial periods in the Quaternary.

In other parts of Australia, and in many parts of Africa, drying out of former soils has not been so severe. This has led to large areas of relict podsols in quite dry climates in the far southern inland of Australia (where temperate rainforest was formerly dominant) and to the formation of torrox soils (a suborder of oxisols) in southern Africa. Here, present climates allow, effectively, the maintenance of the old soils under climates which they could not actually form if one were to start with the parent material on which they developed in the Mesozoic and Paleocene.

Paleosols in this sense are always exceedingly infertile soils, containing available phosphorus levels orders of magnitude lower than in temperate regions with younger soils. Ecological studies have shown that this has forced highly specialised evolution amongst Australian flora to obtain minimal nutrient supplies. The fact that soil formation is simply not occurring makes ecologically sustainable management even more difficult. However, paleosols often contain the most exceptional biodiversity due to the absence of competition.

Duricrust

A duricrust inselberg near Dori, Burkina Faso

Duricrust is a hard layer on or near the surface of soil. Duricrusts can range in thickness from a few millimeters or centimeters to several meters.

It is a general term for a zone of chemical precipitation and hardening formed at or near the surface of sedimentary bodies through pedogenic and (or) non-pedogenic processes. It is typically formed by the accumulation of soluble minerals deposited by mineral-bearing waters that move upward, downward, or laterally by capillary action, commonly assisted in arid settings by evaporation. Minerals often found in duricrust include silica, iron, calcium, and gypsum.

Duricrusts need to be formed in absolute accumulation, therefore they must have a source, transfer and precipitation. Duricrust is often studied during missions to Mars because it may help prove the planet once had more water. Duricrust was found on Mars at the Viking 2 landing site, and a similar structure, nicknamed "Snow Queen," was found under the Phoenix landing site. Phoenix's duricrust was later confirmed to be water-based.

References

- McCarthy, David F. (1982). Essentials of Soil Mechanics and Foundations: Basic Geotechnics (2nd ed.). Reston, Virginia: Reston Publishing. ISBN 9780835917810.
- Gilluly, James; Waters, Aaron Clement; Woodford, Alfred Oswald (1975). Principles of Geology (4th ed.). W.H. Freeman. ISBN 978-0716702696.

- Buol, S. W.; Hole, F. D.; McCracken, R. J. (1973). Soil Genesis and Classification (1st ed.). Ames, Iowa: Iowa State University Press. ISBN 0-8138-1460-X.
- Retallack, G.J. (2008). Soils of the Past: An Introduction to Paleopedology (2nd ed.). John Wiley & Sons. p. 207. ISBN 9780470698167.
- Nyle C. Brady & Ray R. Weil (2009). Elements of the Nature and Properties of Soils (3rd Edition). Prentice Hall. ISBN 9780135014332.
- Paul, E.A. (1997). Soil organic matter in temperate agroecosystems : long-term experiments in North America. Boca Raton: CRC Press. p. 80. ISBN 9780849328022.
- Buol, S.W. (1990). Soil genesis and classification. Ames, Iowa: Iowa State University Press. p. 36. doi:10.1081/E-ESS. ISBN 0-8138-2873-2.
- Drainage Manual: A Guide to Integrating Plant, Soil, and Water Relationships for Drainage of Irrigated Lands. Interior Dept., Bureau of Reclamation. 1993. ISBN 0-16-061623-9.
- Mullins, CE (1997). "Hardsetting". In R Lal, WH Blum, C Valentin, BA Stewart. Methods for assessment of soil degradation. Boca Raton, FL: CRC Press. p. 121. ISBN 9780849374432. Retrieved 18 August 2016.
- Retallack, G. J. (1990). Soils of the past : an introduction to paleopedology. Boston: Unwin Hyman. p. 32. ISBN 978-0-04-445757-2.
- Marsh, William M. (2010). Landscape planning : environmental applications (5th ed.). Hoboken, NJ: Wiley. ISBN 9780470570814.
- Danoff-Burg, James A. "The Terrestrial Influence: Geology and Soils". Earth Institute Center for Environmental Sustainability. Columbia University. Retrieved 27 July 2014.
- College of Tropical Agriculture and Human Resources. "Soil Mineralogy". www.ctahr.hawaii.edu. University of Hawai'i. Retrieved 7 August 2014.

3

Classification of Soil

This chapter will provide a comprehensive overview of all the major types of soil which includes alkali soil, clay and serpentine soil. It also glances upon the process of classification of soil and the factors influencing it. The chapter strives to provide a clear understanding about the types of soil in a systematic yet critical manner.

Soil Classification

Soil classification deals with the systematic categorization of soils based on distinguishing characteristics as well as criteria that dictate choices in use.

Soil types

Overview

Soil classification is a dynamic subject, from the structure of the system itself, to the definitions of classes, and finally in the application in the field. Soil classification can be approached from the perspective of soil as a material and soil as a resource.

Engineering

Engineers, typically geotechnical engineers, classify soils according to their engineering properties as they relate to use for foundation support or building material. Modern engineering classification systems are designed to allow an easy transition from field observations to basic predictions of soil engineering properties and behaviors.

The most common engineering classification system for soils in North America is the Unified Soil Classification System (USCS). The USCS has three major classification groups: (1) coarse-grained soils (e.g. sands and gravels); (2) fine-grained soils (e.g. silts and clays); and (3) highly organic soils (referred to as "peat"). The USCS further subdivides the three major soil classes for clarification. It distinguishes sands from gravels by grain size, and further classifying some as "well-graded" and the rest as "poorly-graded". Silts and clays are distinguished by the soils' Atterberg limits, and separates "high-plasticity" from "low-plasticity" soils as well. Moderately organic soils are considered subdivisions of silts and clays, and are distinguished from inorganic soils by changes in their plasticity properties (and Atterberg limits) on drying. The European soil classification system (ISO 14688) is very similar, differing primarily in coding and in adding an "intermediate-plasticity" classification for silts and clays, and in minor details.

Other engineering soil classification systems in the United States include the AASHTO Soil Classification System and the Modified Burmister system.

A full geotechnical engineering soil description will also include other properties of the soil including color, in-situ moisture content, in-situ strength, and somewhat more detail about the material properties of the soil than is provided by the USCS code. The USCS and additional engineering description is standardized in ASTM D 2487.

Soil Science

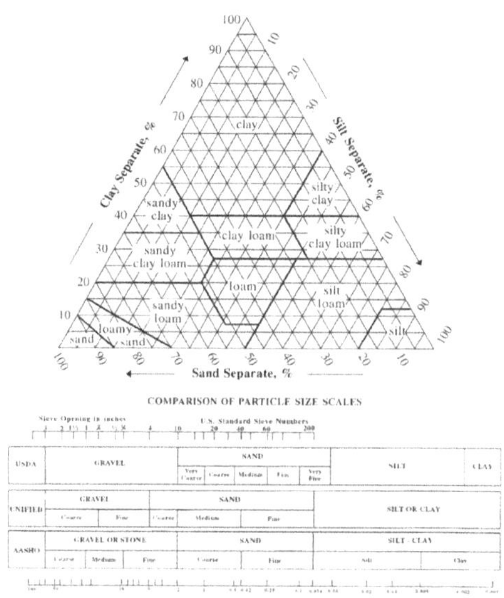

Soil texture triangle showing the USDA classification system based on grain size

For soil resources, experience has shown that a natural system approach to classification, i.e. grouping soils by their intrinsic property (soil morphology), behaviour, or genesis, results in classes that can be interpreted for many diverse uses. Differing concepts of pedogenesis, and differences in the significance of morphological features to various land uses can affect the classification approach. Despite these differences, in a well-constructed system, classification criteria group similar concepts so that interpretations do not vary widely. This is in contrast to a technical system approach to soil classification, where soils are grouped according to their fitness for a specific use and their edaphic characteristics.

Natural system approaches to soil classification, such as the French Soil Reference System (Référentiel pédologique français) are based on presumed soil genesis. Systems have developed, such as USDA soil taxonomy and the World Reference Base for Soil Resources, which use taxonomic criteria involving soil morphology and laboratory tests to inform and refine hierarchical classes.

Another approach is numerical classification, also called ordination, where soil individuals are grouped by multivariate statistical methods such as cluster analysis. This produces natural groupings without requiring any inference about soil genesis.

In soil survey, as practiced in the United States, soil classification usually means criteria based on soil morphology in addition to characteristics developed during soil formation. Criteria are designed to guide choices in land use and soil management. As indicated, this is a hierarchical system that is a hybrid of both *natural* and objective criteria. USDA soil taxonomy provides the core criteria for differentiating soil map units. This is a substantial revision of the 1938 USDA soil taxonomy which was a strictly natural system. Soil taxonomy based soil map units are additionally sorted into classes based on technical classification systems. Land Capability Classes, hydric soil, and prime farmland are some examples.

In addition to scientific soil classification systems, there are also vernacular soil classification systems. Folk taxonomies have been used for millennia, while scientifically based systems are relatively recent developments.

Soil Classifications for Osha

The U.S. Occupational Safety and Health Administration (OSHA) requires the classification of soils to protect workers from injury when working in excavations and trenches. OSHA uses 3 soil classifications plus one for rock, based primarily on strength but also :

- Stable Rock: natural solid mineral matter that can be excavated with vertical sides and remain intact while exposed.
- Type A - cohesive, plastic soils with unconfined compressive strength greater than 1.5 ton per square foot (tsf)(144 kPa), and meeting several other requirements
- Type B - cohesive soils with unconfined compressive strength between 0.5 tsf (48 kPa) and 1.5 tsf (144 kPa), or unstable dry rock, or soils which would otherwise be Type A

- Type C - granular soils or cohesive soils with unconfined compressive strength less than 0.5 tsf (48 kPa) or any submerged or freely seeping soil or adversely bedded soils

Each of the soil classifications has implications for the way the excavation must be made or the protections (sloping, shoring, shielding, etc.) that must be provided to protect workers from collapse of the excavated bank.

Current International System

- Buol, S.W., Southard, R.J., Graham, R.C., and McDaniel, P.A. (2003). Soil Genesis and Classification, 5th Edition. Iowa State Press - Blackwell, Ames, IA.
- Driessen, P., Deckers, J., Spaargaren, O., & Nachtergaele, F. (Eds.). (2001). Lecture notes on the major soils of the world. Rome: FAO.
- FAO. (1998). World Reference Base for Soil Resources. Rome: Food and Agriculture Organization of the United Nations.

Current National Systems

- Agriculture Canada Expert Committee on Soil Survey. (1987). The Canadian system of soil classification (2nd ed.). Ottawa: Canadian Government Publishing Centre.
- Avery, B. W. (1980). Soil classification for England and Wales: higher categories. Cranfield, England: Cranfield University, Soil Survey & Land Research Centre/National Soil Resources Institute.
- Baize, D., & Girard, M. C. (Eds.). (1995). Référentiel pédologique 1995. Paris: Institut National de la Recherche Agronomique.
- Baize, D., & Girard, M. C. (Eds.). (1998). A sound reference base for soils: The "Référentiel Pédologique" (English translation by Hodgson J.M., Eskenazi N.R., & Baize D. ed.). Paris: Institut National de la Recherche Agronomique.
- Hewitt, A. E. (1992). Soil classification in New Zealand: legacy and lessons. Australian Journal of Soil Research, 30, 843-854.
- Isbell, R. F. (1996). The Australian soil classification. Collingwood, Victoria, Australia: CSIRO.
- Soil Classification Working Group. (1991). Soil classification: a taxonomic system for South Africa. Pretoria: Department of Agricultural Development, RSA.

- Soil Survey Staff. (1999). Soil taxonomy: a basic system of soil classification for making and interpreting soil surveys (2nd ed.). Washington, DC: US Department of Agriculture Soil Conservation Service.

Current Technical Systems

Technical soil classification systems focus on representing some specific facet or quality of the soil, rather than a direct pedogenetic classification. Such technical classifications are developed with specific applications in mind, such as soil-water relationships, land quality assessment or geotechnical engineering.

- Boorman, D. B., Hollis, J. M., & Lilly, A. (1995). Hydrology of soil types: a hydrologically-based classification of the soils of the United Kingdom (No. 126): UK Institute of Hydrology.
- Klingebiel, A. A., & Montgomery, P. H. (1961). Land capability classification. Washington, DC: US Government Printing Office.
- Sanchez, P. A., Palm, C. A., & Buol, S. W. (2003). Fertility capability soil classification: a tool to help assess soil quality in the tropics. Geoderma, 114(3-4), 157-185.
- American Society for Testing and Materials, 1985, D 2487-83, Classification of Soils for Engineering Purposes: Annual Book of ASTM Standards. Vol. 04.08, pp 395–408.

Earlier Systems of Historical Interest

- Baldwin, M., Kellogg, C. E., & Thorp, J. (1938). Soil classification. In Soils and men: Yearbook of agriculture (pp. 979–1001). Washington, DC: U.S. Department of Agriculture.
- Simonson, R. W. (1989). Historical aspects of soil survey and soil classification with emphasis on the United States, 1899-1970. Wageningen, NL: International Soil Reference and Information Centre (ISRIC).

Principles

- Butler, B. E. (1980). Soil classification for soil survey. Oxford: Oxford Science Publications.Science, 96,
- Cline, M. G. (1949). Basic principles of soil classification. Soil Science, 67(2), 81-91.
- Cline, M. G. (1963). Logic of the new system of soil classification. Soil 17-22.

- Webster, R. (1968). Fundamental objections to the 7th approximation. Journal of Soil Science, 19, 354-366.
- Terzaghi Karl (1924). Soil Mechanics in Engineering Practice, Wiley-Interscience; 3 Sub-edition (January 1996, ISBN 0-471-08658-4)

Numerical Classification

- McBratney, A. B., & de Gruijter, J. J. (1992). A continuum approach to soil classification by modified fuzzy k-means with extragrades. Journal of Soil Science, 43(1), 159-175.

Alkali Soil

Alkali, or alkaline, soils are clay soils with high pH (> 8.5), a poor soil structure and a low infiltration capacity. Often they have a hard calcareous layer at 0.5 to 1 metre depth. Alkali soils owe their unfavorable physico-chemical properties mainly to the dominating presence of sodium carbonate, which causes the soil to swell and difficult to clarify/settle. They derive their name from the alkali metal group of elements, to which sodium belongs, and that can induce basicity. Sometimes these soils are also referred to as *(alkaline) sodic soils.* Alkaline soils are basic, but not all basic soils are alkaline.

Rice cultivation / paddyfield in alkali soils

Causes

The causes of soil alkalinity are natural or they can be man-made.

1. The natural cause is the presence of soil minerals producing sodium carbonate (Na_2CO_3) and sodium bicarbonate ($NaHCO_3$) upon weathering.

2. Coal-fired boilers / power plants, when using coal or lignite rich in limestone, produce ash containing calcium oxide (CaO). CaO readily dissolves in water to form slaked lime / $Ca(OH)_2$ and carried by rain water to rivers / irrigation water. Lime softening process precipitates Ca and Mg ions / removes hardness in the water and also converts sodium bicarbonates in river water into sodium carbonate. Sodium carbonates (washing soda) further reacts with the remaining Ca and Mg in the water to remove / precipitate the total hardness. Also water-soluble sodium salts present in the ash enhance the sodium content in water. The global coal consumption is 7700 million tons in the year 2011. Thus river water is made devoid of Ca and Mg ions and enhanced Na by coal-fired boilers.

3. Many sodium salts are used in industrial and domestic applications such as Sodium carbonate, Sodium bicarbonate (baking soda), Sodium sulphate, Sodium hydroxide (caustic soda), Sodium hypochlorite (bleaching powder), etc. in huge quantities. These salts are mainly produced from Sodium chloride (common salt). All the sodium in these salts enter into the river / ground water during their production process or consumption enhancing water sodicity. The total global consumption of sodium chloride is 270 million tons in the year 2010. This is nearly equal to the salt load in the mighty Amazon River. Man made sodium salts contribution is nearly 7% of total salt load of all the rivers. Sodium salt load problem aggravates in the downstream of intensively cultivated river basins located in China, India, Egypt, Pakistan, west Asia, Australia, western US, etc. due to accumulation of salts in the remaining water after meeting various transpiration and evaporation losses.

4. Another source of man made sodium salts addition to the agriculture fields / land mass is in the vicinity of the wet cooling towers using sea water to dissipate waste heat generated in various industries located near the sea coast. Huge capacity cooling towers are installed in oil refineries, petrochemical complexes, fertilizer plants, chemical plants, nuclear & thermal power stations, centralized HVAC systems, etc. The drift / fine droplets emitted from the cooling towers contain nearly 6% sodium chloride which would deposit on the vicinity areas. This problem aggravates where the national pollution control norms are not imposed or not implemented to minimize the drift emissions to the best industrial norm for the sea water based wet cooling towers.

5. The man-made cause is the application of softened water in irrigation (surface or ground water) containing relatively high proportion of sodium bicarbonates and less calcium and magnesium.

Agricultural Problems

Alkaline soils are difficult to take into agricultural production. Due to the low infiltration capacity, rain water stagnates on the soil easily and, in dry periods, cultivation is hardly possible without copious irrigated water and good drainage. Agriculture is limited to crops tolerant to surface waterlogging (e.g. rice, grasses) and the productivity is low.

Chemistry

Soil alkalinity is associated with the presence of sodium carbonate (Na_2CO_3) or sodium bicarbonate ($NaHCO_3$) in the soil, either as a result of natural weathering of the soil particles or brought in by irrigation and/or flood water.

- $Na_2CO_3 + H_2O \longleftrightarrow 2NaOH + CO_2$
- $NaHCO_3 \longleftrightarrow NaOH + CO_2$

Sodium carbonate in water can produce carbon dioxide (CO_2), escaping as a gas or absorbed by Algae, and sodium hydroxide (Na^+OH^-), which is alkaline (or rather basic) and gives high pH values (pH>8.5).

- Water (H_2O) is partly dissociated into H_3O^+ (hydronium) and OH^- (hydroxyl) ions. The ion H_3O^+ has a positive electric charge (+) and the hydroxyl group OH^- has a negative charge (–). In pure, neutral water, the concentration of H_3O^+ and OH^- ions equals 10^{-7} eq/l each (respectively $19x10^{-7}$ g/l and $17x10^{-7}$ g/l), very small concentrations.
- 1 eq = 1 equivalent weight corresponds to as many grams of the substance as its formula weight divided by its valence. It is also known as *gram-molecule* or mole per unit of valence. For the hydronium ion (H_3O^+) the formula weight equals 19, and for the hydroxyl group (OH^-) it equals 17.
- In neutral water, the pH, being the negative log value of the H_3O^+ concentration in eq/l, is 7. Similarly, the pOH is also 7. Each unit decrease in pH indicates a tenfold increase of the H_3O^+ concentration. Similarly, each unit increase in pH indicates a tenfold increase of the OH^- concentration.
- In water with dissolved salts, the concentrations of the H_3O^+ y OH^- ions may change, but their sum remains constant, namely 7 + 7 = 14. A pH of 7 therefore corresponds to a pOH of 7, and a pH of 9 with a pOH of 5.
- Formally it is preferred to express the ion concentrations in terms of

chemical activity, but this hardly affects the value of the pH.

- Water with excess H_3O^+ ions is called acid (pH < 7), and water with excess OH^- ions is called alkaline or rather basic (pH > 7). Soil moisture with pH < 4 is called very acid and with pH > 10 very alkaline (basic).

The reaction between Na_2CO_3 and H_2O can be represented as follows:

- $2Na^+ + CO_3^{2-} + 2H^+ + 2OH^- \Rightarrow 2Na^+ + 2OH^- + H_2CO_3$

Carbonic acid H_2CO_3 is unstable and produces H_2O (water) and CO_2 (carbon dioxide gas, escaping into the atmosphere). This explains the remaining alkalinity (or rather basicity) in the form of soluble sodium hydroxide and the high pH or low pOH.

Not all the dissolved sodium carbonate undergoes the above chemical reaction. The remaining sodium carbonate, and hence the presence of CO_3^{2-} ions, causes $CaCO_3$ (which is only slightly soluble) to precipitate as solid calcium carbonate (limestone), because the product of the CO_3^{2-} concentration and the Ca^{2+} concentration exceeds the allowable limit. Hence, the calcium ions Ca^{++} are immobilized.

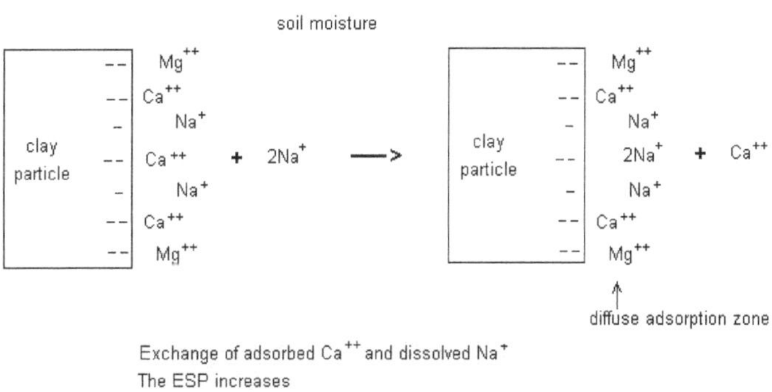

Sodium exchange process between ions adsorbed at the surface of clay particles and those in the soil moisture

The presence of abundant Na^+ ions in the soil solution and the precipitation of Ca^{++} ions as a solid mineral causes the clay particles, which have negative electric charges along their surfaces, to adsorb more Na^+ in the *diffuse adsorption zone* and, in exchange, release previously adsorbed Ca^{++}, by which their *exchangeable sodium percentage* (ESP) is increased as illustrated in the figure.

Na^+ is more mobile and has a smaller electric charge than Ca^{++} so that the thickness of the DAZ increases as more sodium is present. The thickness is also influenced by the total concentration of ions in the soil moisture in the sense that higher concentrations cause the DAZ zone to shrink.

Clay particles with considerable ESP (> 16), in contact with non-saline soil moisture have an expanded DAZ zone and the soil swells (dispersion). The phenomenon results in deterioration of the soil structure, and especially crust formation and compaction of the top layer. Hence the infiltration capacity of the soil and the water availability in the soil is reduced, whereas the surface-water-logging or runoff is increased. Seedling emergence and crop production are badly affected.

Note:

- Under saline conditions, the many ions in the soil solution counteract the swelling of the soil, so that saline soils usually do not have unfavorable physical properties. *Alkaline soils, in principle, are not saline since the alkalinity problem is worse as the salinity is less.*

Alkalinity problems are more pronounced in clay soils than in loamy, silty or sandy soils. The clay soils containing montmorillonite or smectite (swelling clays) are more subject to alkalinity problems than illite or kaolinite clay soils. The reason is that the former types of clay have larger specific surface areas (i.e. the surface area of the soil particles divided by their volume) and higher cation exchange capacity (CEC).

Note:

- Certain clay minerals with almost 100% ESP (i.e. almost fully sodium saturated) are called bentonite, which is used in civil engineering to place impermeable curtains in the soil, e.g. below dams, to prevent seepage of water.

The quality of the irrigation water in relation to the alkalinity hazard is expressed by the following two indexes:

1) The sodium adsorption ratio (SAR,)

The formula for calculating sodium adsorption ratio is:

$$SAR = \frac{[Na+]}{\sqrt{[Ca++/2 + Mg++/2]}} = \frac{\{Na+/23\}}{\sqrt{\{Ca++/40 + Mg++/24\}}}$$

where: [ ] stands for concentration in milliequivalents/liter (briefly meq/l), and { } stands for concentration in mg/l.

It is seen that Mg (Magnesium) is thought to play a similar role as Ca (Calcium).

The SAR should not be much higher than 20 and preferably less than 10;

When the soil has been exposed to water with a certain SAR value for some time, the ESP value tends to become about equal to the SAR value.

2) The residual sodium carbonate (RSC, meq/l,):

The formula for calculating residual sodium carbonate is:

$$RSC = [HCO_3^- + CO_3^=] - [Ca^{++} + Mg^{++}]$$

$$= \{HCO_3^-/61 + CO_3^=/30\} - \{Ca^{++}/20 + Mg^{++}/12\}$$

which must not be much higher than 1 and preferably less than 0.5.

The above expression recognizes the presence of bicarbonates (HCO_3^-), the form in which most carbonates are dissolved.

While calculating SAR and RSC, the water quality present at the root zone of the crop should be considered which would take into account the leaching factor in the field. The partial pressure of dissolved CO_2 at the plants root zone also decides the Calcium present in dissolved form in the field water. USDA follows the adjusted SAR for calculating water sodicity.

Soil Improvement

Alkaline soils with solid $CaCO_3$ can be reclaimed with grass cultures, organic compost, waste hair / feathers, organic garbage, waste paper, rejected lemons/oranges, etc. ensuring the incorporation of much acidifying material (inorganic or organic material) into the soil, and enhancing dissolved Ca in the field water by releasing CO_2 gas. Deep ploughing and incorporating the calcareous subsoil into the top soil also helps.

Many times salts' migration to the top soil takes place from the underground water sources rather than surface sources. Where the underground water table is high and the land is subjected to high solar radiation, ground water oozes to the land surface due to capillary action and gets evaporated leaving the dissolved salts in the top layer of the soil. Where the underground water contains high salts, it leads to acute salinity problem. This problem can be reduced by applying mulch to the land. Using poly-houses or shade netting during summer for cultivating vegetables/crops is also advised to mitigate soil salinity and conserve water / soil moisture. Poly-houses filter the intense summer solar radiation in tropical countries to save the plants from water stress and leaf burns.

Where the ground water quality is not alkaline / saline and ground water table is high, salts build up in the soil can be averted by using the land throughout the year for growing plantation trees / permanent crops with the help of lift irrigation. When the ground water is used at required leaching factor, the salts in the soil would not build up.

Plowing the field soon after cutting the crop is also advised to prevent salt migration to the top soil and conserve the soil moisture during the intense summer months. This is done to break the capillary pores in the soil to prevent water reaching the surface of the soil.

Clay soils in high annual rain fall (more than 100 cm) areas do not generally suffer from high alkalinity as the rain water runoff is able to reduce/leach the soil salts to comfortable levels if proper rain water harvesting methods are followed. In some agricultural areas, the use of subsurface "tile lines" are used to facilitate drainage and leach salts. Continuous Drip irrigation would lead to alkali soils formation in the absence of leaching / drainage water from the field.

It is also possible to reclaim alkaline soils by adding acidifying minerals like pyrite or cheaper alum or Aluminium sulfate.

Alternatively, gypsum (calcium sulfate, $CaSO_4 \cdot 2H_2O$) can also be applied as a source of Ca^{++} ions to replace the sodium at the exchange complex. Gypsum also reacts with sodium carbonate to convert into sodium sulphate which is a neutral salt and does not contribute to high pH. There must be enough natural drainage to the underground, or else an artificial subsurface drainage system must be present, to permit leaching of the excess sodium by percolation of rain and/or irrigation water through the soil profile.

Calcium Chloride is also used to reclaim alkali soils. $CaCl_2$ converts Na_2CO_3 into NaCl precipitating $CaCO_3$. NaCl is drained off by leaching water. Spent acid (HCl, H_2SO_4, etc.) can also be used to reduce the excess Na_2CO_3 in the soil/water.

Where urea is made available cheaply to farmers, it is also used to reduce the soil alkalinity / salinity primarily. The NH_4 (Ammonium) present in urea which is a weak cation releases the strong cation Na from the soil structure into water. Thus alkali soils absorb / consume more urea compared to other soils.

To reclaim the soils completely one needs prohibitively high doses of amendments. Most efforts are therefore directed to improving the top layer only (say the first 10 cm of the soils), as the top layer is most sensitive to deterioration of the soil structure. The treatments, however, need to be repeated in a few (say 5) years' time. Trees / plants follow gravitropism. It is difficult to survive in alkali soils for the trees with deeper rooting system which can be more than 60 meters deep in good non-alkali soils.

It will be important to refrain from irrigation (ground water or surface water) with poor quality water.

One way of reducing sodium carbonate is to cultivate glasswort or saltwort or barilla plants. These plants sequester the sodium carbonate they absorb from alkali soil into their tissues. The ash of these plants contains good quantity of sodium carbonate which can be commercially extracted and used in place of sodium carbonate derived from common salt which is highly energy intensive process. Thus alkali lands deterioration can be checked by cultivating barilla plants which can serve as food source, biomass fuel and raw material for soda ash and potash, etc.

Leaching Saline Sodic Soils

Saline soils are mostly also sodic (the predominant salt is sodium chloride), but they do not have a very high pH nor a poor infiltration rate. Upon leaching they are usually not converted into a (sodic) alkali soil as the Na^+ ions are easily removed. Therefore, saline (sodic) soils mostly do not need gypsum applications for their reclamation.

Clay

Clay is a fine-grained natural rock or soil material that combines one or more clay minerals with traces of metal oxides and organic matter. Clays are plastic due to their water content and become hard, brittle and non–plastic upon drying or firing. Geologic clay deposits are mostly composed of phyllosilicate minerals containing variable amounts of water trapped in the mineral structure. Depending on the soil's content in which it is found, clay can appear in various colours from white to dull gray or brown to deep orange-red.

Gay Head cliffs in Martha's Vineyard consist almost entirely of clay

Electron microscope photograph of smectite clay – magnification 23,500

Clays are distinguished from other fine-grained soils by differences in size and mineralogy. Silts, which are fine-grained soils that do not include clay minerals, tend to have larger particle sizes than clays. There is, however, some overlap in particle size and other physical properties, and many naturally occurring deposits include both silts and clay. The distinction between silt and clay varies by discipline. Geologists and soil scientists usually consider the separation to occur at a particle size of 2 µm (clays being finer than silts), sedimentologists often use 4–5 µm, and colloid chemists use 1 µm. Geotechnical engineers distinguish between silts and clays based on the plasticity properties of the soil, as measured by the soils' Atterberg limits. ISO 14688 grades clay particles as being smaller than 2 µm and silt particles as being larger.

Quaternary clay in Estonia

Formation

Deforestation for clay extraction in Rio de Janeiro, Brazil. The picture is of Morro da Covanca, Jacarepaguá.

Clay minerals typically form over long periods of time as a result of the gradual chemical weathering of rocks, usually silicate-bearing, by low concentrations of carbonic acid and other diluted solvents. These solvents, usually acidic, migrate through the weathering rock after leaching through upper weathered layers. In addition to the weathering process, some clay minerals are formed through hydrothermal activity. There are two types of clay deposits: primary and secondary. Primary clays form as residual deposits in soil and remain at the site of formation. Secondary clays are clays that have been transported from their original location by water erosion and deposited in a new sedimentary deposit. Clay deposits are typically associated with very low energy depositional environments such as large lakes and marine basins.

Grouping

Depending on the academic source, there are three or four main groups of clays: kaolinite, montmorillonite-smectite, illite, and chlorite. Chlorites are not always considered a clay, sometimes being classified as a separate group within the phyllosilicates. There are approximately 30 different types of "pure" clays in these categories, but most "natural" clay deposits are mixtures of these different types, along with other weathered minerals.

Varve (or *varved clay*) is clay with visible annual layers, which are formed by seasonal deposition of those layers and are marked by differences in erosion and organic content. This type of deposit is common in former glacial lakes. When fine sediments are delivered into the calm waters of these glacial lake basins away from the shoreline, they settle to the lake bed. The resulting seasonal layering is preserved in an even distribution of clay sediment banding.

Quick clay is a unique type of marine clay indigenous to the glaciated terrains of Norway, Canada, Northern Ireland, and Sweden. It is a highly sensitive clay, prone to liquefaction, which has been involved in several deadly landslides.

Historical and Modern Uses

Clay layers in a construction site. Dry clay is normally much more stable than sand with regard to excavations.

Clays exhibit plasticity when mixed with water in certain proportions. When dry, clay becomes firm and when fired in a kiln, permanent physical and chemical changes occur. These changes convert the clay into a ceramic material. Because of these properties, clay is used for making pottery, both utilitarian and decorative, and construction products, such as bricks, wall and floor tiles. Different types of clay, when used with different minerals and firing conditions, are used to produce earthenware, stoneware, and porcelain. Prehistoric humans discovered the useful properties of clay. Some of the earliest pottery shards recovered are from central Honshu, Japan. They are associated with the Jomon culture and deposits they were recovered from have been dated to around 14,000 BC.

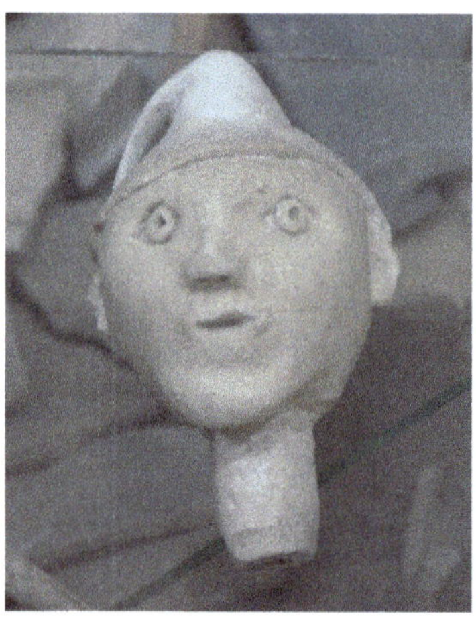

Bottle stopper made of clay, 14th century

Clay tablets were the first known writing medium. Scribes wrote by inscribing them with cuneiform script using a blunt reed called a stylus. Purpose-made clay balls were also used as sling ammunition.

Clays sintered in fire were the first form of ceramic. Bricks, cooking pots, art objects, dishware, and even musical instruments such as the ocarina can all be shaped from clay before being fired. Clay is also used in many industrial processes, such as paper making, cement production, and chemical filtering. Clay is also often used in the manufacture of pipes for smoking tobacco. Until the late 20th century, bentonite clay was widely used as a mold binder in the manufacture of sand castings.

Clay, being relatively impermeable to water, is also used where natural seals are needed, such as in the cores of dams, or as a barrier in landfills against toxic seepage (lining the landfill, preferably in combination with geotextiles).

Recent studies have investigated clay's absorption capacities in various applications, such as the removal of heavy metals from waste water and air purification.

Medicinal Uses

A traditional use of clay as medicine goes back to prehistoric times. An example is Armenian bole, which is used to soothe an upset stomach, similar to the way parrots (and later, humans) in South America originally used it. Kaolin clay and attapulgite have been used as anti-diarrheal medicines.

As a Building Material

Clay is one of the oldest building materials on Earth, among other ancient, naturally-occurring geologic materials such as stone and organic materials like wood. Between one-half and two-thirds of the world's population, in traditional societies as well as developed countries, still live or work in a building made with clay as an essential part of its load-bearing structure. Also a primary ingredient in many natural building techniques, clay is used to create adobe, cob, cordwood, and rammed earth structures and building elements such as wattle and daub, clay plaster, clay render case, clay floors and clay paints and ceramic building material. Clay was used as a mortar in brick chimneys and stone walls where protected from water.

Serpentine Soil

Serpentine outcrop high in the Siskiyou Wilderness of northwest California — here Jeffrey pine dominates the landscape

A serpentine soil is derived from ultramafic rocks, in particular serpentinite, a rock formed by the hydration and metamorphic transformation of ultramafic rock from the Earth's mantle. Serpentinite is composed of the mineral serpentine, but the two terms are often both used to mean the rock, not its mineral composition.

The soils derived from ultramafic bedrock give rise to unusual and sparse associations of edaphic (and often endemic) plants that are tolerant of extreme soil conditions, including:

- low calcium-to-magnesium ratio
- lack of essential nutrients such as nitrogen, potassium, and phosphorus
- high concentrations of nickel and chromium (more common in ultramafic rocks).

The atmosphere is the source of nitrogen, mainly by microbial fixation, in practically all soils. Soils derived from any parent material, including serpentinite, differ greatly in the amounts of plant-available nitrogen in them. Nitrogen spans practically the same range of concentrations in serpentine soils as in most other kinds of soils. The low Ca:Mg ratios is the main plant limiting factor in serpentine soils.

Plants that grow only in serpentine soils are commonly called serpentine endemics.

As Ecologic Indicator

Excellent examples of serpentine soils and the distinctive ecologic communities associated with them have been described in western North America.

For instance, in areas where these ultramafic rocks are patchy, such as the Klamath Mountains of northern California and southwest Oregon, the areas of serpentine soil can be clearly seen as sparsely covered areas or open forest bounded by closed forest on the non-ultramafic soils.

Botany

Areas of serpentine soil are also home to diverse wildflowers, many of which are rare or endangered species such as *Acanthomintha duttonii*, *Pentachaeta bellidiflora*, and *Phlox hirsuta*. In California, shrubs such as leather oak (*Quercus durata*) and coast whiteleaf manzanita (*Arctostaphylos viscida* ssp. *pulchella*) are typical of serpentine soils.

Parent Rock

Serpentine-rich rock or serpentinite has a mottled, greenish-gray color with a waxy feel to it. These rocks form by the reaction of olivine-rich rock, peridotite, with water.

It forms in oceanic crust near the surface of the earth, particularly where water circulates in cooling rock near mid-ocean ridges: masses of the resulting ultramafic rock are found in ophiolites incorporated in continental crust near present and past plate tectonic boundaries.

Distribution

Serpentine soils are widely distributed on Earth, in part mirroring the distribution of ophiolites. Although it covers only about 1 percent of the state's surface, the state rock of California is serpentine. One such area in California is the Edgewood Park and Natural Preserve. Serpentine soils also are present in small but widely distributed areas on the eastern slope of the Appalachian mountains of eastern North America.

Serpentine Endemism and Bioremediation

The unique plants that survive in serpentine soils have been used in the process of phytoremediation, a type of bioremediation. Because those plants developed specialized adaptations to exist in a metal-saturated environment, they have been used to treat environmental problems by taking up certain heavy metals from polluted soil.

Serpentine Barrens

Unlike most ecosystems, in serpentine barrens there is less plant growth closer to a stream, due to toxic minerals in the water.

Serpentine barrens are a unique ecoregion found in parts of the United States in small but widely distributed areas of the Appalachian Mountains and the Coast Ranges of California, Oregon, and Washington. The barrens occur on outcrops of altered ultramafic ophiolites.

They are named for minerals of the serpentine group, resulting in *serpentine soils*, with unusually high concentrations of iron, chromium, nickel and cobalt. Serpentine barrens often consist of grassland or savannas in areas where the climate would normally lead to the growth of forests.

The ecology of serpentine barrens is poorly understood: the evolution of plants adapted to such areas and the relationship between the soil geology and ecology, particularly the ways in which plants handle high concentrations of metals such as nickel, is largely unexplored.

Examples

Soldiers Delight Natural Environmental Area in Baltimore County, Maryland, covers 1,900 acres of serpentine barren. The area has over 38 rare, threatened, or endangered plant species as well as rare insects, rocks and minerals.

Serpentine Barrens in Rock Springs Nature Preserve, Lancaster County, Pennsylvania

Rock Springs Nature Preserve in Lancaster County, Pennsylvania, is a 176-acre (71 ha) property conserved by the Lancaster County Conservancy that is a prime example of a serpentine barren. It was originally a grassland, but fire suppression led to the conversion of the area to forest. This barren contains the rare serpentine aster, as well as a number of rare species of moth and skippers.

In Chester County, Pennsylvania, the Nottingham Park Serpentine Barrens was recommended by UMCES as deserving of National Natural Landmark designation, on numerous grounds, including supporting a number of rare and endemic species, an intact population of pitch pine, and also the site having historic significance.

Soil Gradation

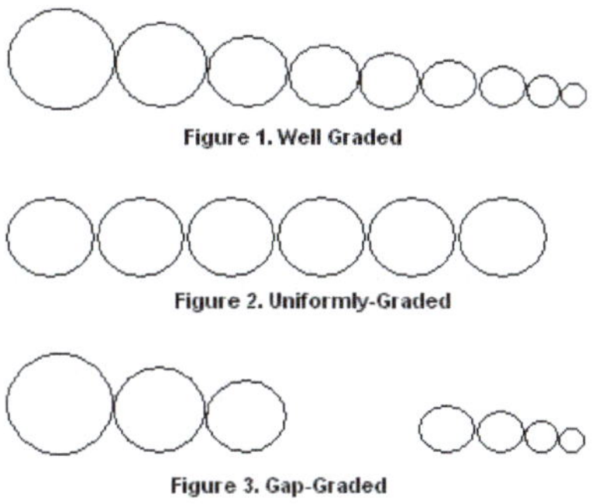

Soil Gradation Categories

Soil gradation is a classification of a coarse-grained soil that ranks the soil based on the different particle sizes contained in the soil. Soil gradation is an important aspect of soil mechanics and geotechnical engineering because it is an indicator of other engineering properties such as compressibility, shear strength, and hydraulic conductivity. In a design, the gradation of the in situ or on site soil often controls the design and ground water drainage of the site. A poorly graded soil will have better drainage than a well graded soil.

Soil is graded as either well graded or poorly graded. ref Dr.B.C PUNMIA- Soil Mechanics and Khanna civil engineering hand book ,Delhi.

Soil gradation is determined by analyzing the results of a sieve analysis or a hydrometer analysis.

The process for grading a soil is in accordance with either the Unified Soil Classification System or the AASHTO Soil Classification System. Gradation of a soil is determined by reading the grain size distribution curve produced from the results of laboratory tests on the soil. Gradation of a soil can also be determined by calculating the coefficient of uniformity, C_u, and the coefficient of curvature, C_c, of the soil and comparing the calculated values with published gradation limits.

Soil Gradations

Soil gradation is a classification of the particle size distribution of a soil. Coarse-grained soils, mainly gravels or sands, are graded as either well graded or poorly graded. Poorly graded soils are further divided into uniformly-graded or gap-graded soils. Fine-grained soils, mainly silts and clays, are classified according to their Atterberg limits.

Well Graded

A *well graded* soil is a soil that contains particles of a wide range of sizes and has a good representation of all sizes from the No. 4 to No. 200 sieves. A well graded gravel is classified as GW while a well graded sand is classified as SW.

Poorly Graded

A *poorly graded* soil is a soil that does not have a good representation of all sizes of particles from the No. 4 to No. 200 sieve. A poorly graded gravel is classified as GP while a poorly graded sand is classified as SP. Poorly graded soils are more susceptible to soil liquefaction than well graded soils. A soil having particles of nearly the same size is known as poorly graded soil. ref-soil mechanics by - Professor B.C punmia and soil engineering by Professor Shamsher Prakash and Khanna civil Engineering hand book ,Delhi, India, IIT Delhi publications. Ref -ASCE ,USA

A *gap-graded* soil is a soil that has an excess or deficiency of certain particle sizes or a soil that has at least one particle size missing. An example of a gap-graded soil is one in which sand of the No. 10 and No. 40 sizes are missing, and all the other sizes are present.

Process of Grading a Soil

The process of grading a soil is in accordance with either the Unified Soil Classification System or the AASHTO Soil Classification System. The steps in grading a soil are data collection, calculating coefficients of uniformity and curvature, and grading the soil based on the grading criteria given in the used soil classification system.

Data Collection

Soil gradation is determined by analyzing the results of a sieve analysis or a hydrometer analysis.

In a sieve analysis, a coarse-grained soil sample is shaken through a series of woven-wire square-mesh sieves. Each sieve has successively smaller openings so particles larger than the size of each sieve are retained on the sieve. The percentage of each soil size is measured by weighing the amount retained on each sieve and comparing the weight to the total weight of the sample. The results of a sieve analysis are plotted as a grain size distribution curve, which is then analyzed to determine the soil gradation of the particular soil.

In a hydrometer analysis, a fine-grained soil sample is left to settle in a viscous fluid. This method is used based on Stoke's Law which relates terminal velocity of fall of a particle in a viscous fluid to the grain diameter and density of the grain in suspension. Grain diameter is calculated from a known distance and time of the fall of the particle. This is used to classify fine-grained soils.

Calculating the Coefficients of Uniformity and Curvature

Calculating the coefficients of uniformity and curvature requires grain diameters. The grain diameter can be found for each percent of the soil passing a particular sieve. This means that if 40% of the sample is retained on the No. 20 sieve then there is 60% passing the No. 20 sieve.

The *coefficient of uniformity*, C_u is a crude shape parameter and is calculated using the following equation:

$$C_u = \frac{D_{60}}{D_{10}}$$

where D_{60} is the grain diameter at 60% passing, and D_{10} is the grain diameter at 10% passing

The *coefficient of curvature*, C_c is a shape parameter and is calculated using the following equation:

$$C_c = \frac{(D_{30})^2}{D_{10} \times D_{60}}$$

where D_{60} is the grain diameter at 60% passing, D_{30} is the grain diameter at 30% passing, and D_{10} is the grain diameter at 10% passing

Once the coefficient of uniformity and the coefficient of curvature have been calculated, they must be compared to published gradation criteria.

Criteria for Grading Soils

The following criteria are in accordance with the Unified Soil Classification System:

For a gravel to be classified as well graded, the following criteria must be met:

$C_u > 4$ & $1 < C_c < 3$

If both of these criteria are not met, the gravel is classified as poorly graded or GP. If both of these criteria are met, the gravel is classified as well graded or GW.

For a sand to be classified as well graded, the following criteria must be met:

$C_u \geq 6$ & $1 < C_c < 3$

If both of these criteria are not met, the sand is classified as poorly graded or SP. If both of these criteria are met, the sand is classified as well graded or SW.

Importance

Soil gradation is very important to geotechnical engineering. It is an indicator of other engineering properties such as compressibility, shear strength, and hydraulic conductivity.

In a design, the gradation of the in situ or on site soil often controls the design and ground water drainage of the site. A poorly graded soil will have better drainage than a well graded soil because there are more void spaces in a poorly graded soil.

When a fill material is being selected for a project such as a highway embankment or earthen dam, the soil gradation is considered. A well graded soil is able to be compacted more than a poorly graded soil. These types of projects may also have gradation requirements that must be met before the soil to be used is accepted.

When options for ground remediation techniques are being selected, the soil gradation is a controlling factor.

References

- Chhabra, R. 1996. Soil Salinity and Water Quality. pp 284. Oxford&IBH Publishing Co. Pvt. Ltd., New Delhi (South Asian edition) and A.A. Balkema Uitgevers BC, Rotterdam (edition elsewhere). ISBN 81-204-1049-1.
- Ehlers, Ernest G. and Blatt, Harvey (1982). 'Petrology, Igneous, Sedimentary, and Metamorphic' San Francisco: W.H. Freeman and Company. ISBN 0-7167-1279-2.
- Alexander, E. B.; Coleman, R. G.; Keeler-Wolf, T.; Harrison, S. (2007). Serpentine Geoecology of Western North America: Geology, Soils, and Vegetation. New York: Oxford University Press. ISBN 0-19-516508-X.
- Dann, Kevin T. (1988). Traces on the Appalachians: A Natural History of Serpentine in Eastern North America. New Brunswick: Rutgers University Press. ISBN 0-8135-1323-5.
- Anderson, Roger C., et al., Savannas, Barrens, and Rock Outcrop Plant Communities of North America, Ch. 19, Cambridge University Press, 1999, ISBN 0-521-57322-X
- Coduto, Donald P. Foundation Design Principles and Practices (2nd Edition). 02nd ed. Upper Saddle River: Prentice Hall, 2000. Print. ISBN 0-13-589706-8
- "University College London Geology on Campus: Clays". Earth Sciences department. University College London. Retrieved 10 January 2016.
- I.P. Abrol, J.S.P. Yadav and F.I. Massoud. "Salt-Affected Soils and their Management, refer para 4.7". Retrieved 23 December 2012.

4

Various Aspects of Soil

Soil depicts various characteristics and properties which are significant for a complete understanding of soil science. Soil type, Soil pH, Soil fertility, etc. are some of the important topics related to soil science which are rigorously examined in this chapter. It unfolds these significant aspects in a crucial manner.

Soil pH

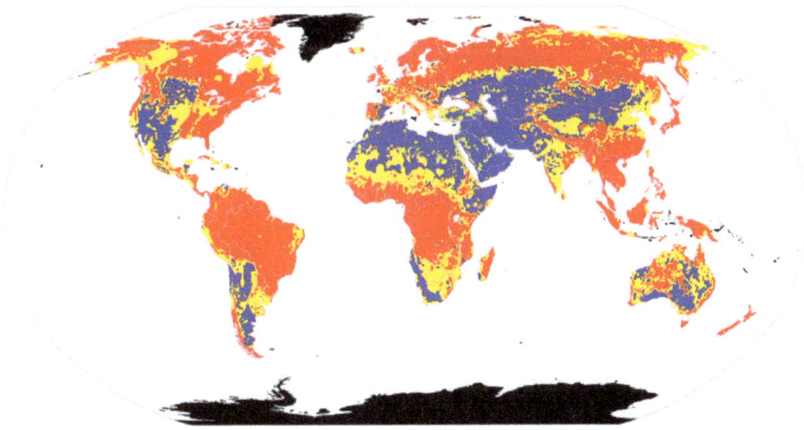

Global variation in soil pH. Red = acidic soil. Yellow = neutral soil. Blue = alkaline soil. Black = no data.

The soil pH is a measure of the acidity or alkalinity in soils. pH is defined as the negative logarithm (base 10) of the activity of hydronium ions (H+ or, more precisely, H3O+aq) in a solution. In water, it normally ranges from -1 to 14, with 7 being neutral. A pH below 7 is acidic and above 7 is alkaline. Soil pH is considered a master variable in soils as it controls many chemical processes that take place. It specifically affects plant nutrient availability by controlling the chemical forms of the nutrient. The optimum pH range for most plants is between 5.5 and 7.0, however many plants have adapted to thrive at pH values outside this range.

Classification of Soil pH Ranges

The United States Department of Agriculture Natural Resources Conservation Service, formerly Soil Conservation Service classifies soil pH ranges as follows:

Denomination	pH range
Ultra acid	< 3.5
Extremely acid	3.5–4.4
Very strongly acid	4.5–5.0
Strongly acid	5.1–5.5
Moderately acid	5.6–6.0
Slightly acid	6.1–6.5
Neutral	6.6–7.3
Slightly alkaline	7.4–7.8
Moderately alkaline	7.9–8.4
Strongly alkaline	8.5–9.0
Very strongly alkaline	> 9.0

Sources Of Soil pH

Sources of Acidity

Acidity in soils comes from H^+ and Al^{3+} ions in the soil solution and sorbed to soil surfaces. While pH is the measure of H^+ in solution, Al^{3+} is important in acid soils because between pH 4 and 6, Al^{3+} reacts with water (H_2O) forming $AlOH^{2+}$, and $Al(OH)_2^+$, releasing extra H^+ ions. Every Al^{3+} ion can create 3 H^+ ions. Many other processes contribute to the formation of acid soils including rainfall, fertilizer use, plant root activity and the weathering of primary and secondary soil minerals. Acid soils can also be caused by pollutants such as acid rain and mine spoilings.

- Rainfall: Acid soils are most often found in areas of high rainfall. Excess rainfall leaches base cation from the soil, increasing the percentage of Al^{3+} and H^+ relative to other cations. Additionally, rainwater has a slightly acidic pH of 5.7 due to a reaction with CO_2 in the atmosphere that forms carbonic acid.
- Fertilizer use: Ammonium (NH_4^+) fertilizers react in the soil in a process called nitrification to form nitrate (NO_3^-), and in the process release H^+ ions.
- Plant root activity: Plants take up nutrients in the form of ions (NO_3^-, NH_4^+, Ca^{2+}, $H_2PO_4^-$, etc.), and often, they take up more cations than anions. However plants must maintain a neutral charge in their roots. In order to compensate for the extra positive charge, they will release H^+ ions from the root. Some plants will also exude organic acids into the soil to acidify the zone around their roots to help solubilize metal nutrients that are insoluble at neutral pH, such as iron (Fe).
- Weathering of minerals: Both primary and secondary minerals that compose soil contain Al. As these minerals weather, some components such as Mg, Ca, and K, are taken up by plants, others such as Si are leached from the soil, but

due to chemical properties, Fe and Al remain in the soil profile. Highly weathered soils are often characterized by having high concentrations of Fe and Al oxides.

- Acid rain: When atmospheric water reacts with sulfur and nitrogen compounds that result from industrial processes, the result can be the formation of sulfuric and nitric acid in rainwater. However the amount of acidity that is deposited in rainwater is much less, on average, than that created through agricultural activities.
- Mine spoil: Severely acidic conditions can form in soils near mine spoils due to the oxidation of pyrite.
- Potential acid sulfate soils naturally formed in waterlogged coastal and estuarine environments can become highly acidic when drained or excavated.
- Decomposition of organic matter by microorganisms releases CO_2 which when mixed with soil water can form carbonic acid (H_2CO_3).

Sources of Alkalinity

Alkaline soils have a high saturation of base cations (K^+, Ca^{2+}, Mg^{2+} and Na^+). This is due to an accumulation of soluble salts which are classified as either saline soil, sodic soil, saline-sodic soil or alkaline soil. All saline and sodic soils have high salt concentrations, with saline soils being dominated by calcium and magnesium salts and sodic soils being dominated by sodium. Alkaline soils are characterized by the presence of carbonates. Soil in areas with limestone near the surface are alkaline from the calcium carbonate in limestone constantly mixing with the soil. Groundwater sources in these areas contain dissolved limestone.

Effect of Soil pH on Plant Growth

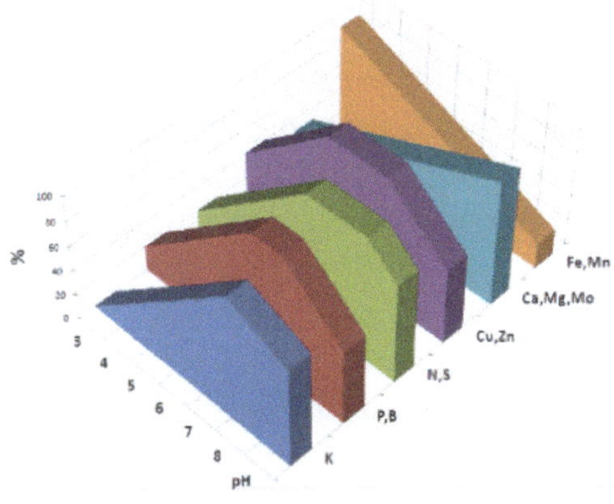

Nutrient availability in relation to soil pH

Acid Affected Soils

Plants grown in acid soils can experience a variety of symptoms including aluminium (Al), hydrogen (H), and/or manganese (Mn) toxicity, as well as nutrient deficiencies of calcium (Ca) and magnesium (Mg).

Aluminium toxicity is the most widespread problem in acid soils. Aluminium is present in all soils, but dissolved Al^{3+} is toxic to plants; Al^{3+} is most soluble at low pH, above pH 5.2 little Al is in soluble form in most soils. Aluminium is not a plant nutrient, and as such, is not actively taken up by the plants, but enters plant roots passively through osmosis. Aluminium inhibits root growth; lateral roots and root tips become thickened and roots lack fine branching; root tips may turn brown. In the root, Al has been shown to interfere with many physiological processes including the uptake and transport of calcium and other essential nutrients, cell division, cell wall formation, and enzyme activity.

Below pH 4, H^+ ions themselves damage root cell membranes.

In soils with high content of manganese-containing minerals, Mn toxicity can become a problem at pH 5.6 and lower. Manganese, like aluminium, becomes increasingly soluble as pH drops, and Mn toxicity symptoms can be seen at pH levels below 5.6. Manganese is an essential plant nutrient, so plants transport Mn into leaves. Classic symptoms of Mn toxicity are crinkling or cupping of leaves.

Nutrient Availability in Relation to Soil ph

Nutrients needed in large amounts by plants are referred to as macrontrents and iclude nitrogen (N), phosphorus (P), potassium (K), calcium (Ca), magnesium (Mg) and sulfur (S). Elements that plants need in trace amounts are called trace nutrients or micronutrients. Trace nutrients are not major components of plant tissue but are essential for growth. They include iron (Fe), manganese (Mn), zinc (Zn), coper (Cu), cobalt (Co), molybdenum (Mo), and boron (B). Both macronutrient and micronutrient availability are affected by soil pH. In slightly to moderately alkaline soils, molybdenum and macronutrient (except for phosphorus) availability is increased, but P, Fe, Mn, Zn Cu, and Co levels are reduced and may adversely affect plant growth. In acidic soils, micronutrient availability (except for Mo and Bo) is increased. Nitrogen is supplied as ammonium (NH4) or nitrate (NO3) by nitrogen fixation or fertilizer amendments, and dissolved N will have the highest concentrations in soil with pH 6–8. Concentrations of available N are less sensitive to pH than concentration of available P. In order for P to be available for plants, soil pH needs to be in the range 6.0 and 7.5. If pH is lower than 6, P starts forming insoluble compounds with iron (Fe) and aluminium (Al) and if pH is higher than 7.5 P starts forming insoluble compounds with calcium (Ca). Most nutrient deficiencies can be avoided between a pH range of 5.5 to 6.5, provided that soil minerals and organic matter contain the essential nutrients to begin with.

Water Availability in Relation to Soil pH

Determining pH

Methods of determining pH include:

- Observation of soil profile: Certain profile characteristics can be indicators of either acid, saline, or sodic conditions. Strongly acidic soils often have poor incorporation of the organic surface layer with the underlying mineral layer. The mineral horizons are distinctively layered in many cases, with a pale eluvial (E) horizon beneath the organic surface; this E is underlain by a darker B horizon in a classic podzol horizon sequence. This is a very rough gauge of acidity as there is no correlation between thickness of the E and soil pH. E horizons a few feet thick in Florida usually have pH just above 5 (merely "strongly acid") while E horizons a few inches thick in New England are "extremely acid" with pH readings of 4.5 or below. In the southern Blue Ridge Mountains there are "ultra acid" soils, pH below 3.5, which have no E horizon. Presence of a caliche layer indicates the presence of calcium carbonates, which are present in alkaline conditions. Also, columnar structure can be an indicator of sodic condition.

- Observation of predominant flora. Calcifuge plants (those that prefer an acidic soil) include *Erica*, *Rhododendron* and nearly all other Ericaceae species, many birch (*Betula*), foxglove (*Digitalis*), gorse (*Ulex* spp.), and Scots Pine (*Pinus sylvestris*). Calcicole (lime loving) plants include ash trees (*Fraxinus* spp.), honeysuckle (*Lonicera*), *Buddleja*, dogwoods (*Cornus* spp.), lilac (*Syringa*) and *Clematis* species.

- Use of an inexpensive pH testing kit, where in a small sample of soil is mixed with indicator solution which changes colour according to the acidity/alkalinity.

- Use of litmus paper. A small sample of soil is mixed with distilled water, into which a strip of litmus paper is inserted. If the soil is acidic the paper turns red, if alkaline, blue.

- Use of a commercially available electronic pH meter, in which a rod is inserted into moistened soil and measures the concentration of hydrogen ions.

Examples of Plant pH Preferences

- pH 4.5–5.0: Ericaceae (azalea, bilberry, blueberry, cranberry; heather), hydrangea for blue, (less acidic for pink), liquidambar or sweet gum, orchid, pin oak

- pH 5.0–5.5: Boronia, daphne; Ericaceae: (camellia, heather[, rhododendron), ferns, iris, orchids, parsley, conifers (e.g., pine); Poaceae: (maize, millet, rye, oat), radish; Solanales: (potato, sweet potato); Bromeliaceae (pineapple).

- pH 5.5–6.0: Asteraceae: (aster, endive); Brassicaceae: (brussels sprout, kohlrabi), carrot; Cucurbitales: (begonia, chayote or choko); Fabaceae: (bean, crimson clover, peanut, soybean), petunia, rhubarb, violet, most bulbs (canna, daffodil, jonquil), larkspur, primrose
- pH 6.0–6.5 antirrhinum or snapdragon, Brassicaceae: (broccoli, cabbage, candytuft, cauliflower, turnip, wallflower); Cucurbitaceae: (cucumber, pumpkin, squash); Fabaceae: (pea, red clover, white clover), gladiolus, Iceland poppy; Rosales: (cannabis, rose, strawberry); Solanaceae: (eggplant or aubergine, tomato), sweet corn; Violaceae: (pansy, viola), zinnia or zinnea
- pH 6.5–7.0: Amaranthaceae: (beet, spinach); Apiaceae: (celery, parsnip); Asparagales: (asparagus, onion); Asteraceae: (chrysanthemum, dahlia, lettuce), carnation; Fabaceae: (alfalfa, sweet pea), melon, stock, tulip
- pH 7.1–8.0 lilac

Changing Soil pH

Increasing ph of Acidic Soil

The most common amendment to increase soil pH is lime ($CaCO_3$ or $MgCO_3$), usually in the form of finely ground agricultural lime. The amount of lime needed to change pH is determined by the mesh size of the lime (how finely it is ground)and the buffering capacity of the soil. A high mesh size (60–100) indicates a finely ground lime, that will react quickly with soil acidity. Buffering capacity of soils is a function of a soils cation exchange capacity, which is in turn determined by the clay content of the soil, the type of clay and the amount of organic matter present. Soils with high clay content, particularly shrink–swell clay, will have a higher buffering capacity than soils with little clay. Soils with high organic matter will also have a higher buffering capacity than those with low organic matter. Soils with high buffering capacity require a greater amount of lime to be added than a soil with a lower buffering capacity for the same incremental change in pH. Other amendments that can be used to increase the pH of soil include wood ash, industrial CaO (burnt lime), and oyster shells. White firewood ash includes metal salts which are important for processes requiring ions such as Na^+ (sodium), K^+ (potassium), Ca^{2+} (calcium), which may or may not be good for the select flora, but decreases the acidic quality of soil. These products increase the pH of soils through the reaction of CO_3^{2-} with H^+ to produce CO_2 and H_2O. Calcium silicate neutralizes active acidity in the soil by removing free hydrogen ions, thereby increasing pH. As its silicate anion captures H^+ ions (raising the pH), it forms monosilicic acid (H_4SiO_4), a neutral solute.

Decreasing ph of Alkaline Soil

- Iron sulfates or aluminium sulfate as well as elemental sulfur (S) reduce pH through the formation of sulfuric acid.

- Urea, urea phosphate, ammonium nitrate, ammonium phosphates, ammonium sulfate and monopotassium phosphate fertilizers.
- organic matter in the form of plant litter, compost, and manure will decrease soil pH through the decomposition process. Certain acid organic matter such as pine needles, pine sawdust and acid peat are effective at reducing pH.

Soil Salinity

Soil salinity is the salt content in the soil; the process of increasing the salt content is known as salinization. Salts occur naturally within soils and water. Salination can be caused by natural processes such as mineral weathering or by the gradual withdrawal of an ocean. It can also come about through artificial processes such as irrigation.

Visibly salt-affected soils on rangeland in Colorado. Salts dissolved from the soil accumulate at the soil surface and are deposited on the ground and at the base of the fence post.

Natural Occurrence

Salts are a natural component in soils and water. The ions responsible for salination are: Na^+, K^+, Ca^{2+}, Mg^{2+} and Cl^-. As the Na^+ (sodium) predominates, soils can become sodic. Sodic soils present particular challenges because they tend to have very poor structure which limits or prevents water infiltration and drainage.

Over long periods of time, as soil minerals weather and release salts, these salts are flushed or leached out of the soil by drainage water in areas with sufficient precipitation. In addition to mineral weathering, salts are also deposited via dust and precipitation. In dry regions salts may accumulate, leading to naturally saline soils. This is the case, for example, in large parts of Australia. Human practices can increase the salinity of soils by the addition of salts in irrigation water. Proper irrigation management can prevent salt accumulation by providing adequate drainage water to leach added salts from the soil. Disrupting drainage patterns that provide leaching can also result in salt accumulations. An example of this occurred in Egypt in 1970 when the Aswan High Dam was built. The change in the level of ground water before the construction had enabled soil erosion, which led to high concentration of salts in the water table. After the construction, the continuous high level of the water table led to the salination of the arable land.

Dry Land Salinity

Salinity in drylands can occur when the water table is between two and three metres from the surface of the soil. The salts from the groundwater are raised by capillary action to the surface of the soil. This occurs when groundwater is saline (which is true in many areas), and is favored by land use practices allowing more rainwater to enter the aquifer than it could accommodate. For example, the clearing of trees for agriculture is a major reason for dryland salinity in some areas, since deep rooting of trees has been replaced by shallow rooting of annual crops.

Salinity Due to Irrigation

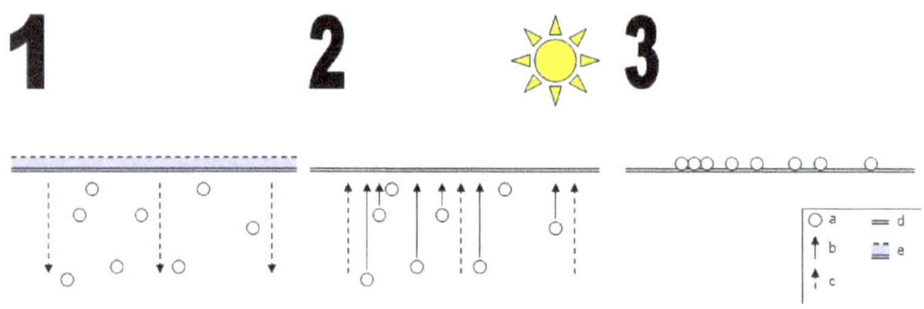

Rain or irrigation, in the absence of leaching, can bring salts to the surface by capillary action

Salinity from irrigation can occur over time wherever irrigation occurs, since almost all water (even natural rainfall) contains some dissolved salts. When the plants use the water, the salts are left behind in the soil and eventually begin to accumulate. Since soil salinity makes it more difficult for plants to absorb soil moisture, these salts must be leached out of the plant root zone by applying additional water. This water in excess of plant needs is called the leaching fraction. Salination from irrigation water is also greatly increased by poor drainage and use of saline water for irrigating agricultural crops.

Salinity in urban areas often results from the combination of irrigation and groundwater processes. Irrigation is also now common in cities (gardens and recreation areas).

Consequences of Salinity

The consequences of salinity are

- detrimental effects on plant growth and yield

- damage to infrastructure (roads, bricks, corrosion of pipes and cables)
- reduction of water quality for users, sedimentation problems
- soil erosion ultimately, when crops are too strongly affected by the amounts of salts.

Salinity is an important land degradation problem. Soil salinity can be reduced by leaching soluble salts out of soil with excess irrigation water. Soil salinity control involves watertable control and flushing in combination with tile drainage or another form of subsurface drainage. A comprehensive treatment of soil salinity is available from the United Nations Food and Agriculture Organization.

High levels of soil salinity can be tolerated if salt-tolerant plants are grown. Sensitive crops lose their vigor already in slightly saline soils, most crops are negatively affected by (moderately) saline soils, and only salinity resistant crops thrive in severely saline soils. The University of Wyoming and the Government of Alberta report data on the salt tolerance of plants.

Regions Affected

From the FAO/UNESCO Soil Map of the World the following salinised areas can be derived.

Region	Area (10^6 ha)
Africa	69.5
Near and Middle East	53.1
Asia and Far East	19.5
Latin America	59.4
Australia	84.7
North America	16.0
Europe	20.7

Soil Salinity Control

Soil salinity control relates to controlling the problem of soil salinity and reclaiming salinized agricultural land.

The aim of soil salinity control is to prevent soil degradation by salination and reclaim already salty (saline) soils. Soil reclamation is also called soil improvement, rehabilitation, remediation, recuperation, or amelioration.

The primary man-made cause of salinization is irrigation. River water or groundwater used in irrigation contains salts, which remain behind in the soil after the water has evaporated.

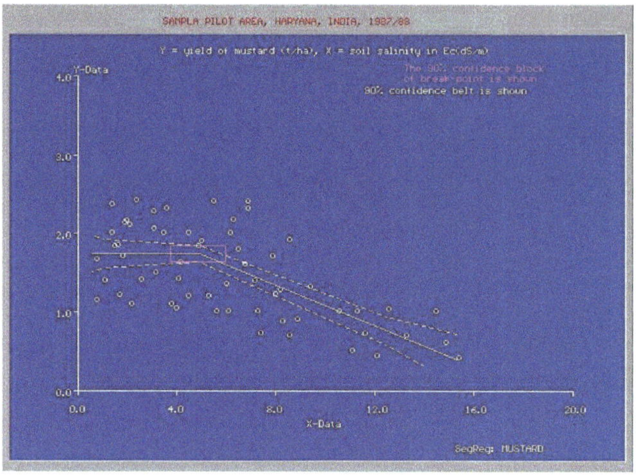

SegReg program: yield of mustard (colza) and soil salinity

The primary method of controlling soil salinity is to permit 10-20% of the irrigation water to leach the soil, be drained and discharged through an appropriate drainage system. The salt concentration of the drainage water is normally 5 to 10 times higher than that of the irrigation water, thus salt export matches salt import and it will not accumulate.

The Soil Salinity Problem

Salty (saline) soils are soils that have a high salt content. The predominant salt is normally sodium chloride (NaCl, "table salt"). Saline soils are therefore also *sodic soils* but there may be sodic soils that are not saline, but alkaline.

World Soil Salt Degradation

This damage is an average of 2,000 hectares of irrigated land in arid and semi-arid areas daily for more than 20 years across 75 countries (each week the world loses an area larger than Manhattan)...To feed the world's anticipated nine billion people by 2050, and with little new productive land available, it's a case of all lands needed on deck.*—principal author Manzoor Qadir, Assistant Director, Water and Human Development, at UN University's Canadian-based Institute for Water, Environment and Health*

According to a study by UN University, about 62 million hectares (20%) of the world's irrigated lands are affected, up from 45 million hectares in the early 1990s. In the Indo-Gangetic Plain, home to over 10% of the world's population, crop yield losses for wheat, rice, sugarcane and cotton grown on salt-affected lands could be 40%, 45%, 48%, and 63%, respectively.

Salty soils are a common feature and an environmental problem in irrigated lands in arid and semi-arid regions, resulting in poor or little crop production. The problems are often associated with high water tables, caused by a lack of natural subsurface drainage to the underground. Poor subsurface drainage may be caused by insufficient transport capacity of the aquifer or because water cannot exit the aquifer, for instance if the aquifer is situated in a topographical depression.

Worldwide, the major factor in the development of saline soils is a lack of precipitation. Most naturally saline soils are found in (semi)arid regions and climates of the earth.

Primary Cause

Irrigated saline land with poor crop stand

For example, assuming irrigation water with a low salt concentration of 0.3 g/l (equal to 0.3 kg/m^3 corresponding to an electric conductivity of about 0.5 FdS/m) and a modest annual supply of irrigation water of 10,000 m^3/ha (almost 3 mm/day) brings 3,000 kg salt/ha each year. In the absence of sufficient natural drainage (as in waterlogged soils) and without a proper leaching and drainage program to remove salts, this would lead to a high soil salinity and reduced crop yields in the long run.

Much of the water used in irrigation has a higher salt content than in this example, which is compounded by the fact that many irrigation projects use a far greater annual supply of water. Sugar cane, for example, needs about 20,000 m^3/ha of water per year. As a result, irrigated areas often receive more than 3,000 kg/ha of salt per year and some receive as much as 10,000 kg/ha/year.

Secondary Cause

The secondary cause of salinization is waterlogging in irrigated land. Irrigation causes changes to the natural water balance of irrigated lands. Large quantities of water in irrigation projects are not consumed by plants and must go somewhere. In irrigation projects it is impossible to achieve 100% irrigation efficiency where all the irrigation water is consumed by the plants. The maximum attainable irrigation efficiency is about 70% but usually it is less than 60%. This means that minimum 30%, but usually more than 40% of the irrigation water is not evaporated and it must go somewhere.

Most of the water lost this way is stored underground which can change the original hydrology of local aquifers considerably. Many aquifers cannot absorb and transport these quantities of water and so the water table rises leading to water logging.

Waterlogging causes three problems:

- The shallow water table and lack of oxygenation of the root zone reduces the yield of most crops
- It leads to an accumulation of salts brought in with the irrigation water as their removal through the aquifer is blocked
- With the upward seepage of groundwater more salts are brought into the soil and the salination is aggravated

Aquifer conditions in irrigated land and the groundwater flow have an important role in soil salinization, as illustrated here :

- Illustration of the influence of aquifer conditions on soil salinization in irrigated land

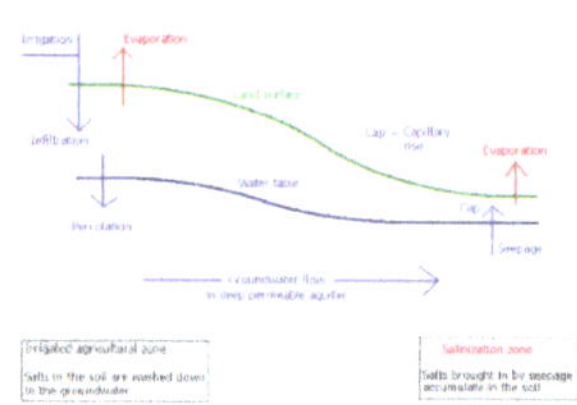

Soil salinization in the lower parts of undulating land with a good aquifer

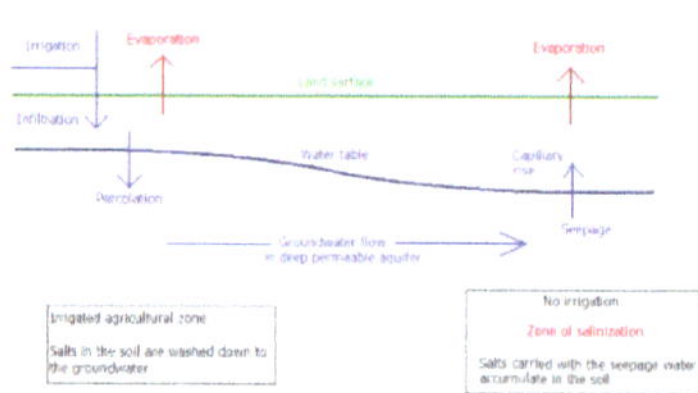

Soil salinization in the unirrigated parts of flat land with a good aquifer

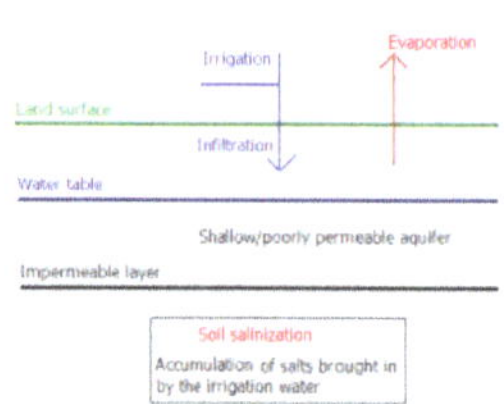

Soil salinization in irrigated flat land without an aquifer

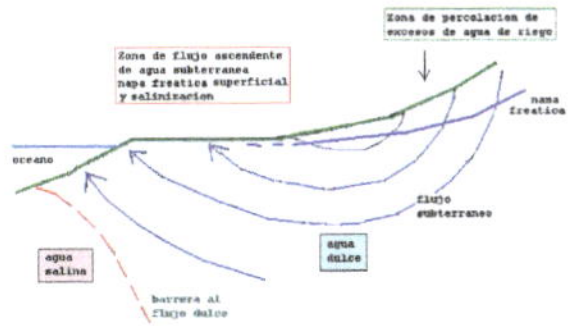

Soil salinization in a coastal delta from irrigation higher up

Salt Affected Area

Normally, the salinization of agricultural land affects a considerable area of irrigation projects, on the order of 20 to 30%. When the agriculture in such a fraction of the land is abandoned, a new salt and water balance is attained, a new equilibrium is reached, and the situation becomes stable.

In India alone, thousands of square kilometres have been severely salinized. China and Pakistan do not lag much behind (perhaps China has even more salt affected land than India). A regional distribution of the 3,230,000 km² of saline land worldwide is shown in the following table derived from the FAO/UNESCO Soil Map of the World.

Region	Area (10^6ha)
Australia	84.7
Africa	69.5
Latin America	59.4
Near and Middle East	53.1
Europe	20.7
Asia and Far East	19.5
Northern America	16.0

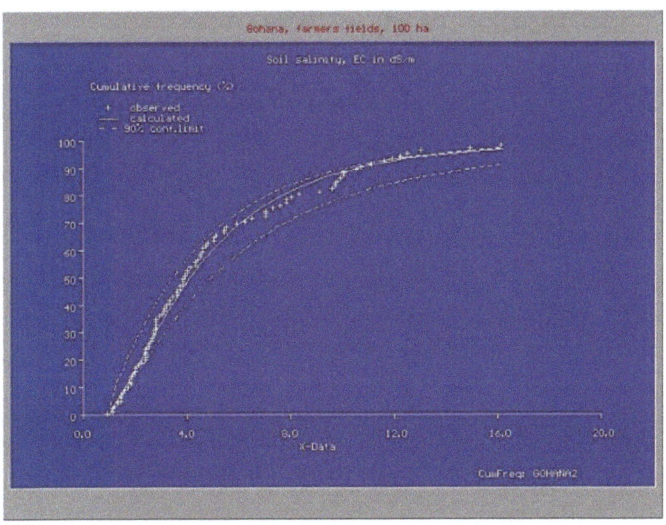

CumFreq program: spatial variation of soil salinity

Spatial Variation

Although the principles of the processes of salinization are fairly easy to understand, it is more difficult to explain why certain parts of the land suffer from the problems and other parts do not, or to predict accurately which part of the land will fall victim. The main reason for this is the variation of natural conditions in time and space, the usually uneven distribution of the irrigation water, and the seasonal or yearly changes of agricultural practices. Only in lands with undulating topography is the prediction simple: the depressional areas will degrade the most.

The preparation of salt and water balances for distinguishable sub-areas in the irrigation project, or the use of agro-hydro-salinity models, can be helpful in explaining or predicting the extent and severity of the problems.

Diagnosis

Measurement

Soil salinity is measured as the salt concentration of the soil solution in tems of g/l or electric conductivity (EC) in dS/m. The relation between these two units is about 5/3 : y g/l => 5y/3 dS/m. Seawater may have a salt concentration of 30 g/l (3%) and an EC of 50 dS/m.

The standard for the determination of soil salinity is from an extract of a saturated paste of the soil, and the EC is then written as ECe. The extract is obtained by centrifugation. The salinity can more easily be measured, without centrifugation, in a 2:1 or 5:1 water:soil mixture (in terms of g water per g dry soil) than from a saturated paste. The relation between ECe and $EC_{2:1}$ is about 4, hence : ECe = 4 $EC_{1:2}$.

Classification

Soils are considered saline when the ECe > 4. When 4 < ECe < 8, the soil is called slightly saline, when 8 < ECe < 16 it is called (moderately) saline, and when ECe > 16 severely saline.

Crop Tolerance

Sensitive crops lose their vigor already in slightly saline soils, most crops are negatively affected by (moderately) saline soils, and only salinity resistant crops thrive in severely saline soils. The University of Wyoming and the Government of Alberta report data on the salt tolerance of plants.

Principles of Salinity Control

Drainage is the primary method of controlling soil salinity. The system should permit a small fraction of the irrigation water (about 10 to 20 percent, the drainage or leaching fraction) to be drained and discharged out of the irrigation project.

In irrigated areas where salinity is stable, the salt concentration of the drainage water is normally 5 to 10 times higher than that of the irrigation water. Salt export matches salt import and salt will not accumulate.

When reclaiming already salinized soils, the salt concentration of the drainage water will initially be much higher than that of the irrigation water (for example 50 times higher). Salt export will greatly exceed salt import, so that with the same drainage fraction a rapid desalinization occurs. After one or two years, the soil salinity is decreased so much, that the salinity of the drainage water has come down to a normal value and a new, favorable, equilibrium is reached.

In regions with pronounced dry and wet seasons, the drainage system may be operated in the wet season only, and closed during the dry season. This practice of checked or controlled drainage saves irrigation water.

The discharge of salty drainage water may pose *environmental problems* to downstream areas. The environmental hazards must be considered very carefully and, if necessary mitigating measures must be taken. If possible, the drainage must be limited to wet seasons only, when the salty effluent inflicts the least harm.

Drainage Systems

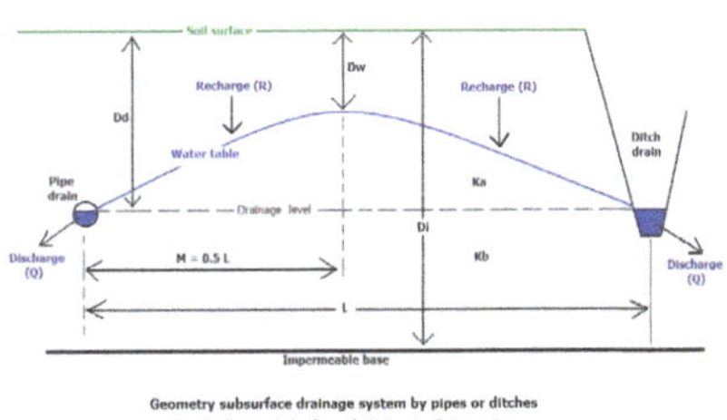

Parameters of a horizontal drainage system

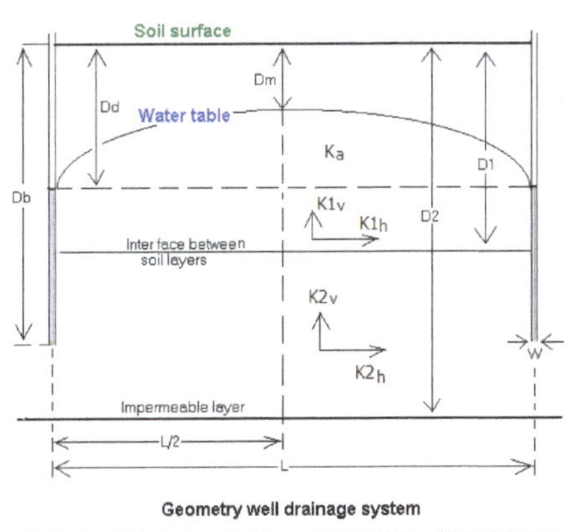

Parameters of a vertical drainage system

Land drainage for soil salinity control is usually by *horizontal drainage system* (figure left), but *vertical systems* (figure right) are also employed.

The drainage system designed to evacuate salty water also lowers the water table. To reduce the cost of the system, the lowering must be reduced to a minimum. The highest permissible level of the water table (or the shallowest permissible depth) depends on the irrigation and agricultural practices and kind of crops.

In many cases a seasonal average water table depth of 0.6 to 0.8 m is deep enough. This means that the water table may occasionally be less than 0.6 m (say 0.2 m just after an irrigation or a rain storm). This automatically implies that, in other occasions, the water table will be deeper than 0.8 m (say 1.2 m). The fluctuation of the water table helps in the breathing function of the soil while the expulsion of carbon dioxide (CO_2) produced by the plant roots and the inhalation of fresh oxygen (O_2) is promoted.

The establishing of a not too deep water table offers the additional advantage that excessive field irrigation is discouraged, as the crop yield would be negatively affected by the resulting elevated water table, and irrigation water may be saved.

The statements made above on the optimum depth of the water table are very general, because in some instances the required water table may be still shallower than indicated (for example in rice paddies), while in other instances it must be considerably deeper (for example in some orchards). The establishment of the optimum depth of the water table is in the realm of agricultural drainage criteria.

Soil Leaching

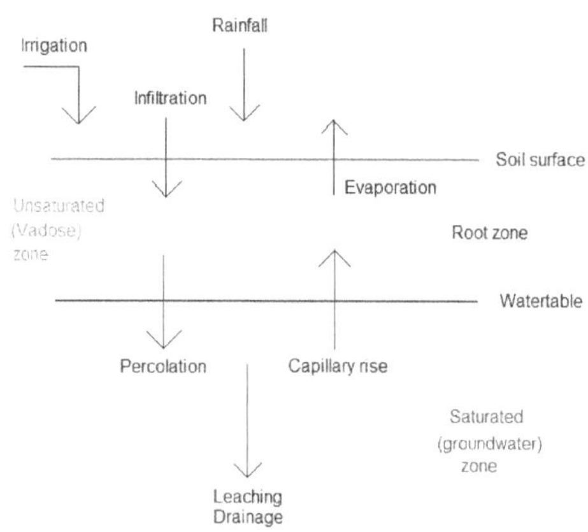

Water balance factors in the soil

The vadose zone of the soil below the soil surface and the watertable is subject to four main hydrological inflow and outflow factors:

- Infiltration of rain and irrigation water (Irr) into the soil through the soil surface (Inf) :

- Inf = Rain + Irr
- Evaporation of soil water through plants and directly into the air through the soil surface (Evap)
- Percolation of water from the unsaturated zone soil into the groundwater through the watertable (Perc)
- Capillary rise of groundwater moving by capillary suction forces into the unsaturated zone(Cap)

In steady state (i.e. the amount of water stored in the unsaturated zone does not change in the long run) the water balance of the unsaturated zone reads: Inflow = Outflow, thus:

- Inf + Cap = Evap + Perc or :
- Irr + Rain + Cap = Evap + Perc

and the *salt balance* is

- Irr.Ci + Cap.Cc = Evap.Fc.Ce + Perc.Cp + Ss

where Ci is the salt concentration of the irrigation water, Cc is the salt concentration of the capillary rise, equal to the salt concentration of the upper part of the groundwater body, Fc is the fraction of the total evaporation transpired by plants, Ce is the salt concentration of the water taken up by the plant roots, Cp is the salt concentration of the percolation water, and Ss is the increase of salt storage in the unsaturated soil. This assumes that the rainfall contains no salts. Only along the coast this may not be true. Further it is assumed that no runoff or surface drainage occurs. The amount of removed by plants (Evap.Fc.Ce) is usually negligibly small: Evap.Fc.Ce = 0

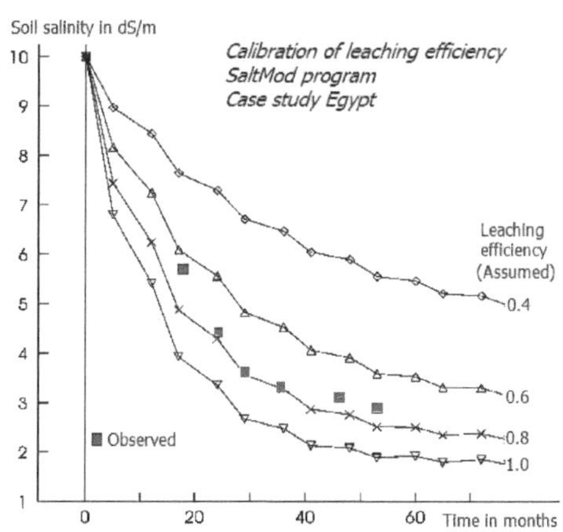

Leaching curves, calibrating leaching efficiency

The salt concentration Cp can be taken as a part of the salt concentration of the soil in the unsaturated zone (Cu) giving: Cp=Le.Cu, where Le is the leaching efficiency. The leaching efficiency is often in the order of 0.7 to 0.8, but in poorly structured, heavy clay soils it may be less. In the Leziria Grande polder in the delta of the Tagus river in Portugal it was found that the leaching efficiency was only 0.15. Assuming that one wishes to avoid the soil salinity to increase and maintain the soil salinity Cu at a desired level Cd we have: Ss = 0, Cu = Cd and Cp = Le.Cd. Hence the salt balance can be simplified to:

- Perc.Le.Cd = Irr.Ci + Cap.Cc

Setting the amount percolation water required to fulfill this salt balance equal to Lr (the *leaching requirement*) it is found that:

- Lr = (Irr.Ci + Cap.Cc) / Le.Cd .

Substituting herein Irr = Evap + Perc – Rain – Cap and re-arranging gives :

- Lr = [(Evap–Rain).Ci + Cap(Cc–Ci)] / (Le.Cd – Ci)

With this the irrigation and drainage requirements for salinity control can be computed too. In irrigation projects in (semi)arid zones and climates it is important to check the leaching requirement, whereby the *field irrigation efficiency* (indicating the fraction of irrigation water percolating to the underground) is to be taken into account. The desired soil salinity level Cd depends on the crop tolerance to salt. The University of Wyoming, USA, and the Government of Alberta, Canada, report crop tolerance data.

Strip Cropping: An Alternative

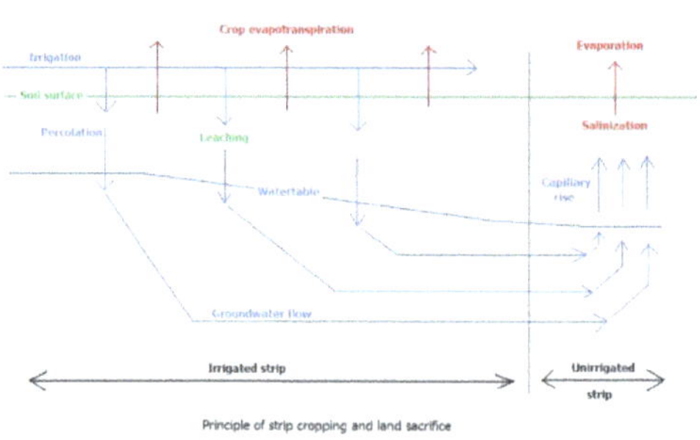

Hydrological principles of *strip cropping* to control the depth of the water table and the soil salinity

In irrigated lands with scarce water resources suffering from drainage (high water table) and soil salinity problems, strip cropping is sometimes practiced with strips of land where every other strip is irrigated while the strips in between are left permanently fallow.

Owing to the water application in the irrigated strips they have a higher watertable which induces flow of groundwater to the unirrigated strips. This flow functions as subsurface drainage for the irrigated strips, whereby the water table is maintained at a not-too-shallow depth, leaching of the soil is possible, and the soil salinity can be controlled at an acceptably low level.

In the unirrigated (sacrificial) strips the soil is dry and the groundwater comes up by capillary rise and evaporates leaving the salts behind, so that here the soil salinizes. Nevertheless, they can have some use for livestock, sowing salinity resistant grasses or weeds. Moreover, useful salt resistant trees can be planted like Casuarina, Eucalyptus or Atriplex, keeping in mind that the trees have deep rooting systems and the salinity of the wet subsoil is less than of the topsoil. In these ways wind erosion can be controlled. The unirrigated strips can also be used for salt harvesting.

Soil Salinity Models

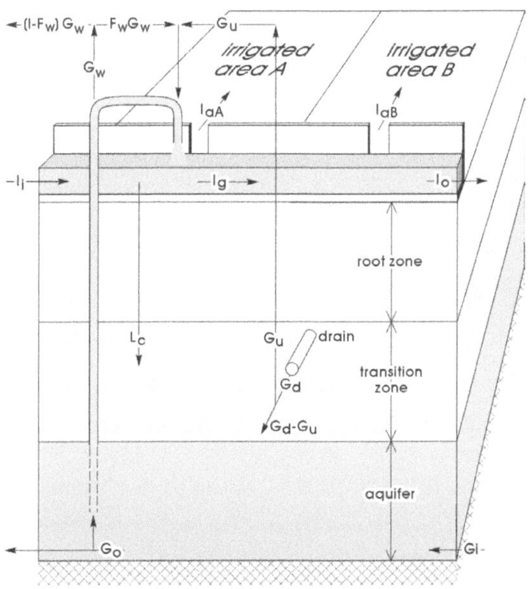

SaltMod components

The majority of the computer models available for water and solute transport in the soil (e.g. SWAP, DrainMod-S, UnSatChem, and Hydrus) are based on Richard's differential equation for the movement of water in unsaturated soil in combination with Fick's differential convection–diffusion equation for advection and dispersion of salts.

The models require input of soil characteristics like the relations between variable unsaturated soil moisture content, water tension, water retention curve, unsaturated hydraulic conductivity, dispersivity and diffusivity. These relations vary to a great extent from place to place and from time to time and are not easy to measure. Further, the models are difficult to calibrate under farmer's field conditions because the soil salinity here is spatially very variable. The models use short time steps and need at least a daily, if not an hourly, data base of hydrological phenomena. Altogether this makes model application to a fairly large project the job of a team of specialists with ample facilities.

Simpler models, like SaltMod, based on monthly or seasonal water and soil balances and an empirical capillary rise function, are also available. They are useful for long-term salinity predictions in relation to irrigation and drainage practices.

LeachMod, using the SaltMod principles, helps in analyzing leaching experiments in which the soil salinity was monitored in various root zone layers while the model will optimize the value of the leaching efficiency of each layer so that a fit is obtained of observed with simulated soil salinity values.

Spatial variations owing to variations in topography can be simulated and predicted using salinity cum groundwater models, like SahysMod.

Soil Fertility

Soil fertility refers to the ability of a soil to sustain plant growth, i.e. to provide plant habitat and result in sustained and consistent yields of high quality. A fertile soil has the following properties:

- The ability to supply essential plant nutrients and soil water in adequate amounts and proportions for plant growth and reproduction; and
- The absence of toxic substances which may inhibit plant growth.

The following properties contribute to soil fertility in most situations:

- Sufficient soil depth for adequate root growth and water retention;
- Good internal drainage, allowing sufficient aeration for optimal root growth (although some plants, such as rice, tolerate waterlogging);
- Topsoil with sufficient soil organic matter for healthy soil structure and soil moisture retention;
- Soil pH in the range 5.5 to 7.0 (suitable for most plants but some prefer or tolerate more acid or alkaline conditions);
- Adequate concentrations of essential plant nutrients in plant-available forms;
- Presence of a range of microorganisms that support plant growth.

In lands used for agriculture and other human activities, maintenance of soil fertility typically requires the use of soil conservation practices. This is because soil erosion and other forms of soil degradation generally result in a decline in quality with respect to one or more of the aspects indicated above.

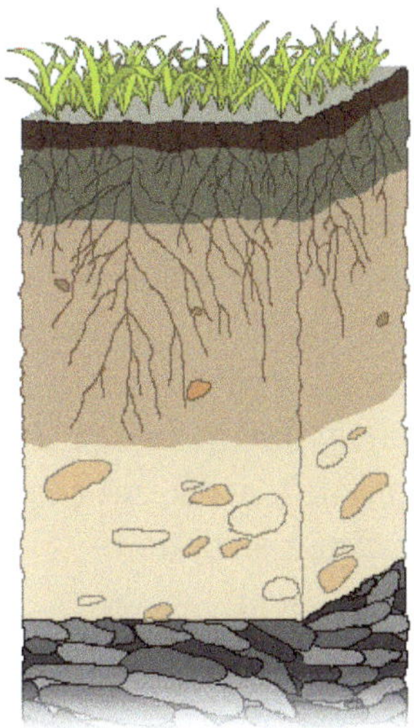

Soil scientists use the capital letters O, A, B, C, and E to identify the master horizons, and lowercase letters for distinctions of these horizons. Most soils have three major horizons—the surface horizon (A), the subsoil (B), and the substratum (C). Some soils have an organic horizon (O) on the surface, but this horizon can also be buried. The master horizon, E, is used for subsurface horizons that have a significant loss of minerals (eluviation). Hard bedrock, which is not soil, uses the letter R.

Soil Fertilization

Bioavailable phosphorus is the element in soil that is most often lacking. Nitrogen and potassium are also needed in substantial amounts. For this reason these three elements are always identified on a commercial fertilizer analysis. For example, a 10-10-15 fertilizer has 10 percent nitrogen, 10 percent (P_2O_5) available phosphorus and 15 percent (K_2O) water-soluble potassium. Sulfur is the fourth element that may be identified in a commercial analysis—e.g. 21-0-0-24 which would contain 21% nitrogen and 24% sulfate.

Inorganic fertilizers are generally less expensive and have higher concentrations of nutrients than organic fertilizers. Also, since nitrogen, phosphorus and potassium generally must be in the inorganic forms to be taken up by plants, inorganic fertilizers are generally immediately bioavailable to plants without modification. However, some have criticized the use of inorganic fertilizers, claiming that the water-soluble nitrogen doesn't provide for the long-term needs of the plant and creates water pollution. Slow-release fertilizers may reduce leaching loss of nutrients and may make the nutrients that they provide available over a longer period of time.

Soil fertility is a complex process that involves the constant cycling of nutrients between organic and inorganic forms. As plant material and animal wastes are decomposed by micro-organisms, they release inorganic nutrients to the soil solution, a process referred to as mineralization. Those nutrients may then undergo further transformations which may be aided or enabled by soil micro-organisms. Like plants, many micro-organisms require or preferentially use inorganic forms of nitrogen, phosphorus or potassium and will compete with plants for these nutrients, tying up the nutrients in microbial biomass, a process often called immobilization. The balance between immobilization and mineralization processes depends on the balance and availability of major nutrients and organic carbon to soil microorganisms. Natural processes such as lightning strikes may fix atmospheric nitrogen by converting it to (NO_2). Denitrification may occur under anaerobic conditions (flooding) in the presence of denitrifying bacteria. The cations, primarily phosphate and potash, as well as many micronutrients are held in relatively strong bonds with the negatively charged portions of the soil in a process known as cation-exchange capacity.

In 2008 the cost of phosphorus as fertilizer more than doubled, while the price of rock phosphate as base commodity rose eight-fold. Recently the term peak phosphorus has been coined, due to the limited occurrence of rock phosphate in the world.

Light and CO_2 Limitations

Photosynthesis is the process whereby plants use light energy to drive chemical reactions which convert CO_2 into sugars. As such, all plants require access to both light and carbon dioxide to produce energy, grow and reproduce.

While typically limited by nitrogen, phosphorus and potassium, low levels of carbon dioxide can also act as a limiting factor on plant growth. Peer-reviewed and published scientific studies have shown that increasing CO_2 is highly effective at promoting plant growth up to levels over 300 ppm. Further increases in CO_2 can, to a very small degree, continue to increase net photosynthetic output.

Since higher levels of CO_2 have only a minimal impact on photosynthetic output at present levels (presently around 400 ppm and increasing), we should not consider plant growth to be limited by carbon dioxide. Other biochemical limitations, such as soil organic content, nitrogen in the soil, phosphorus and potassium, are far more often in short supply. As such, neither commercial nor scientific communities look to air fertilization as an effective or economic method of increasing production in agriculture or natural ecosystems. Furthermore, since microbial decomposition occurs faster under warmer temperatures, higher levels of CO_2 (which is one of the causes of unusually fast climate change) should be expected to increase the rate at which nutrients are leached out of soils and may have a negative impact on soil fertility.

Soil Depletion

Soil depletion occurs when the components which contribute to fertility are removed and not replaced, and the conditions which support soil's fertility are not maintained. This leads to poor crop yields. In agriculture, depletion can be due to excessively intense cultivation and inadequate soil management.

Soil fertility can be severely challenged when land use changes rapidly. For example, in Colonial New England, colonists made a number of decisions that depleted the soils, including: allowing herd animals to wander freely, not replenishing soils with manure, and a sequence of events that led to erosion. William Cronon wrote that "...the long-term effect was to put those soils in jeopardy. The removal of the forest, the increase in destructive floods, the soil compaction and close-cropping wrought by grazing animals, plowing--all served to increase erosion."

One of the most widespread occurrences of soil depletion as of 2008 is in tropical zones where nutrient content of soils is low. The combined effects of growing population densities, large-scale industrial logging, slash-and-burn agriculture and ranching, and other factors, have in some places depleted soils through rapid and almost total nutrient removal.

Topsoil depletion occurs when the nutrient-rich organic topsoil, which takes hundreds to thousands of years to build up under natural conditions, is eroded or depleted of its original organic material. Historically, many past civilizations' collapses can be attributed to the depletion of the topsoil. Since the beginning of agricultural production in the Great Plains of North America in the 1880s, about one-half of its topsoil has disappeared.

Depletion may occur through a variety of other effects, including overtillage (which damages soil structure), underuse of nutrient inputs which leads to mining of the soil nutrient bank, and salinization of soil.

Irrigation Water Effects

The quality of irrigation water is very important to maintain soil fertility and tilth, and for using more soil depth by the plants. When soil is irrigated with high alkaline water, unwanted sodium salts build up in the soil which would make soil draining capacity very poor. So plant roots can not penetrate deep in to the soil for optimum growth in Alkali soils. When soil is irrigated with low pH / acidic water, the useful salts (Ca, Mg, K, P, S, etc.) are removed by draining water from the acidic soil and in addition unwanted aluminium and manganese salts to the plants are dissolved from the soil impeding plant growth. When soil is irrigated with high salinity water or sufficient water is not draining out from the irrigated soil, the soil would convert in to saline soil or lose its fertility. Saline water enhance the turgor pressure or osmotic pressure requirement which impedes the off take of water and nutrients by the plant roots.

Top soil loss takes place in alkali soils due to erosion by rain water surface flows or drainage as they form colloids (fine mud) in contact with water. Plants absorb water-soluble inorganic salts only from the soil for their growth. Soil as such does not lose fertility just by growing crops but it lose its fertility due to accumulation of unwanted and depletion of wanted inorganic salts from the soil by improper irrigation and acid rain water (quantity and quality of water). The fertility of many soils which are not suitable for plant growth can be enhanced many times gradually by providing adequate irrigation water of suitable quality and good drainage from the soil.

Global Distribution

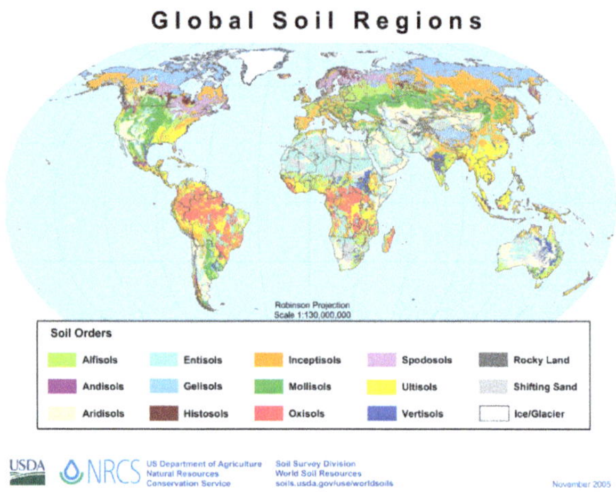

Global distribution of soil types of the USDA soil taxonomy system. Mollisols, shown here in dark green, are a good (though not the only) indicator of high soil fertility. They coincide to a large extent with the world's major grain producing areas like the North American Prairie States, the Pampa and Gran Chaco of South America and the Ukraine-to-Central Asia Black Earth belt.

Soil Type

In terms of soil texture, soil type usually refers to the different sizes of mineral particles in a particular sample. Soil is made up in part of finely ground rock particles, grouped according to size as sand and silt in addition to clay, organic material such as decomposed plant matter.

Each component, and their size, play an important role. For example, the largest particles, sand, determine aeration and drainage characteristics, while the tiniest, sub-microscopic clay particles, are chemically active, binding with water and plant nutrients. The ratio of these sizes determines soil type: clay, loam, clay-loam, silt-loam, and so on.

In addition to the mineral composition of soil, humus (organic material) also plays an important role in soil characteristics and fertility for plant life. Soil may be mixed with larger aggregate, such as pebbles or gravel. Not all types of soil are permeable, such as pure clay.

There are many recognized soil classifications, both international and national

Cation Exchange Capacity

In soil science, cation-exchange capacity or CEC is the number of exchangeable cations per dry weight that a soil is capable of holding, at a given pH value, and available for exchange with the soil water solution. CEC is used as a measure of soil fertility, nutrient retention capacity, and the capacity to protect groundwater from cation contamination. It is expressed as milliequivalent of hydrogen per 100 g of dry soil (meq+/100g), or the SI unit centi-mol per kg (cmol+/kg). The numeric values are the same in any system of units.

Clay and humus have electrostatic surface charges that attract and hold ions. The holding capacity of clay varies with the type of clay. Humus has a CEC that is two to three times that of the best clay.

One way to increase the CEC of a soil is to enhance the formation of humus.

In general, the higher the CEC, the higher the fertility of that soil.

Calculation of CEC

The CEC is the number of positive charges (cations) that a representative sample of soil can hold. It is usually described as the number of hydrogen ions (H^+) necessary to fill the soil cation holding sites per 100 grams of dry soil. Alternatively, an equivalent amount of another cation (Al^{3+} or Ca^{2+}) can be used in the measure. In soil science, an equivalent is defined by the number of charges in terms of a given number of hydrogen ions. As hydrogen ions have only one positive charge (H^+), this makes calculations relatively simple. An equivalent of Al^{3+} that could be held would amount to one third as many of those ions, and Ca^{2+} would have half as many ions.

Translation from meq/100g to an applicable unit, like lb/acre of available nutrients, can be made via calculation, that considers the atomic weight, the ion's valence, and by estimating the soil depth and its density. Mengel gives the following correspondence for 1 meq/100g:

Calcium, 400 lb/acre

Magnesium, 240 lb/acre

Potassium, 780 lb/acre

Ammonium, 360 lb/acre

Base Saturation

Closely related to cation-exchange capacity is the base saturation, which is the fraction of exchangeable cations that are base cations (Ca, Mg, K and Na). It can be expressed as a percentage, and called *percent base saturation*. The higher the amount of exchangeable base cations, the more acidity can be neutralised in the short time perspective. Thus, a soil with high cation-exchange capacity takes longer time to acidify (as well as to recover from an acidified status) than a soil with a low cation-exchange capacity (assuming similar base saturations). A rain, with its load of acidic hydrogen ions, upon a soil that has a high CEC will be quickly returned (buffered) to its original pH in a very short time. A rain on a low CEC soil, such as in the Amazon Basin with its acid soils, will not be restored and the pH will drop sharply and remain there for a relatively long time.

The base-cation saturation ratio (BCSR) is a method of interpreting soil test results that is widely used in sustainable agriculture, supported by the National Sustainable Agriculture Information Service (ATTRA) and claimed to be successfully in use on over a million acres (4,000 km^2) of farmland worldwide.

pH and CEC

For many soils, the CEC is dependent upon the pH of the soil. This is due mostly to the Hofmeister series (lyotrophic series), which describes the relative strength of various cations' adsorption to colloids, and is generally as follows:

$Al^{3+} > H^+ > Ca^{2+} > Mg^{2+} > K^+ = NH_4^+ > Na^+$

As soil acidity increases (pH decreases), more H^+ ions are attached to the colloids. They have pushed the other cations from the colloids and into the soil water solution. Inversely, when soils become more basic (pH increases), the available cations in solution decreases because there are fewer H^+ ions to push cations into the soil solution from the colloids (CEC increases).

Various Colloids and Soils' CEC

The CEC of various soils and soil constituents varies greatly.

Cation exchange capacity for soils; soil textures; soil colloids		
Soil	**State**	**CEC meq/100 g**
Charlotte fine sand	Florida	1.0
Ruston fine sandy loam	Texas	1.9
Glouchester loam	New Jersey	11.9
Grundy silt loam	Illinois	26.3
Gleason clay loam	California	31.6

Susquehanna clay loam	Alabama	34.3
Davie mucky fine sand	Florida	100.8
Sands	–	1–5
Fine sandy loams	–	5–10
Loams and silt loams	–	5–15
Clay loams	–	15–30
Clays	–	over 30
Sesquioxides	–	0–3
Kaolinite	–	3–15
Illite	–	25–40
Montmorillonite	–	60–100
Vermiculite (similar to illite)	–	80–150
Humus	–	100–300

Aluminium Ions and CEC

Many heavily leached or oxidized soils, especially in the wet tropics, have a high concentration of Al^{3+} occupying the soil colloids cation exchange sites. Since aluminium is toxic in high quantities for most plants, there are certain advantages to this. Due to the relatively high adsorption rate of aluminium to soil colloids, it will be taken out of the soil, hence the plant cannot be adversely affected by it. On the other hand, because it has three positive charges, it takes up a large amount of charge on a colloid. For example, Al^{3+} fills the same space as three NH_4^+ ions. As a result, the ammonium is left in the soil water solution where it can be washed away by a heavy rain. This makes many aluminium heavy soils relatively infertile. There is no easy way to remove aluminium ions from the soil colloid and free the CEC for other ions.

Organic Matter

Organic materials in soil increase the CEC through an increase in available negative charges. As such, organic matter build-up in soil usually positively impacts soil fertility. However, organic matter CEC is heavily impacted by soil acidity as acidity causes many organic compounds to release ions to the soil solution.

Anion Exchange Capacity

Similar to the CEC, the anion exchange capacity is a measurement of the positive charges in soils affecting the amount of negative charges which a soil can absorb. There are relatively few anions that are restrictive in agriculture, but they are important, such as sulfur or phosphorus. The anion lyotrophic series is:

$H_2PO_4^- > SO_4^{-2} > NO_3^- > Cl^-$

Converse to CEC, AEC generally will increase when pH drops and decrease when pH rises.

Laboratory Determination

There are two standardised International Soil Reference and Information Centre methods for determining CEC:

- extraction with ammonium acetate; and
- the silver-thiourea method (one-step centrifugal extraction).

There exist slightly conflicting ideas on which mechanisms to include in the term, "cation exchange", in soil chemistry. From a theoretical point of view, one should distinguish cation exchange from ligand exchange, and exchange of diffuse layer adsorbed cations. On the other hand, from a practical point of view, e.g. in forest and agricultural management, what is important is the soils' ability to replace one cation with another rather than the exact mechanism by which this replacement occurs. What is included in the term, "cation exchange", in soil science thus varies with the scientific context.

Standard Values

Kaolinite	3–15
Halloysite 2H_2O	5–10
Halloysite 4H_2O	40–50
Montmorillonite-group	70–100
Illite	10–40
Vermiculite	100–150
Chlorite	10–40
Glauconite	11–20+
Palygorskite-group	20–30
Allophane	~70

These are the values reported by Carroll (1959) for the cation-exchange capacity of minerals in meq/100g at pH of 7.

Mud

A pair of muddy boots.

Mud is a mixture of water and any combination of soil, silt, and clay, and it usually forms after rainfall or near water sources. Ancient mud deposits harden over geological time to form sedimentary rock such as shale or mudstone (generally called lutites). When geological deposits of mud are formed in estuaries the resultant layers are termed bay muds.

Building and Construction

Mud plastered home in Pakistan

Adhesive

In the construction industry, mud is a semi-fluid material that can be used to coat, seal, or adhere materials. Depending on the composition of the mud, it can be referred by many different names, including slurry, mortar, plaster, stucco, and concrete.

Material

Mud, cob, adobe, clay, and many other names are historically used synonymously to mean a mixture of subsoil and water possibly with the addition of stones, gravel, straw, lime, and/or bitumen. This material was used a variety of ways to build walls, roofs and floors.

Mud Brick

Mud house in Amran, Yemen

Mud can be made into mud bricks,also called adobe, by mixing mud with water, placing the mixture into moulds and then allowing it to dry in open air. Straw is sometimes used as a binder within the bricks, as it adds a support lattice. When the brick would otherwise break, the straw will redistribute the force throughout the brick, decreasing the chance of breakage. Such buildings must be protected from groundwater, usually by building upon a masonry, fired brick, rock or rubble foundation, and also from wind-driven rain in damp climates, usually by deep roof overhangs. In extremely dry climates a well drained flat roof may be protected with a well-prepared (puddled) and properly maintained dried mud coating, viable as the mud will expand when moistened and so become more water resistant. Adobe mudbricks were commonly used by the Pueblo Indians to build their homes and other necessary structures.

Fired Brick

Mud that is mostly clay, or a mixture of clay and sand may be used for ceramics, of which one form is the common fired brick. Fired brick are more durable but consume much more energy to produce.

Stabilized Mud

Stabilized mud (earth, soil) is mud which has had a binder such as cement or bitumen added. Examples are mudcrete, landcrete, and soil cement.

Art and Pottery

Pottery is made by forming a clay body into objects of a required shape and heating them to high temperatures in a kiln which removes all the water from the clay, which induces reactions that lead to permanent changes including increasing their strength and hardening and setting their shape. A clay body can be decorated before or after firing. Prior to some shaping processes, clay must be prepared. Kneading helps to ensure an even moisture content throughout the body. Air trapped within the clay body needs to be removed. This is called de-airing and can be accomplished by a machine called a vacuum pug or manually by wedging. Wedging can also help produce an even moisture content. Once a clay body has been kneaded and de-aired or wedged, it is shaped by a variety of techniques. After shaping it is dried and then fired.

In ceramics, the making of liquid mud (called Slip) is a stage in the process of refinement of the materials, since larger particles will settle from the liquid.

Habitat

Land

Mud can provide a home for numerous types of animals, including varieties of worms, frogs, snails, clams, and crayfish. Other animals, such as hippopotamuses, pigs, rhinoceroses, water buffalo and elephants, bathe in mud in order to cool off and protect themselves from the sun.

Marine Life

Mud plays an important role in the marine ecosystem. The activities of burrowing animals and fish have a dramatic churning effect on muddy seabeds. This allows the exchange and cycling of oxygen, nutrients, and minerals between water and sediment.

Below the surface, the burrows of some species form intricate lattice-like networks and may penetrate a meter or more downwards. This means that the burrowed mud is a productive habitat, providing food and shelter for a wide range of mud-dwellers and other animals that forage in and over the mud.

Problems

A landslide on a railroad.

Mud can pose problems for motor traffic when moisture is present, because every vehicle function that changes direction or speed relies on friction between the tires and the road surface, so a layer of mud on the surface of the road or tires can cause the vehicle to hydroplane.

Heavy rainfall, snowmelt, or high levels of ground water may trigger a movement of soil or sediments, possibly causing mudslides, landslides, avalanches, or sinkholes.

Mudslides in volcanic terrain (called Lahars) occur after eruptions as rain remobilizes loose ash deposits.

Mudslides are also common in the western United States during El Niño years due to prolonged rainfall.

As Food

Mud and dirt can be consumed accidentally during sports and other outdoor activities. This has led to dysphemisms for poor-tasting food such as "tastes like dirt", based on the experience of getting mud, dirt, etc. in one's mouth.

There also exist children's recipes for "mud", which is generally a chocolate or cornstarch-based sludge used more for visual appeal than actual taste. Never does this confectionery mud actually contain real mud.

Recreation

Mud Bath

A mud bath is a bath of mud, commonly from areas where hot spring water can combine with volcanic ash. Mud baths have existed for thousands of years, and can be found now in high-end spas in many countries of the world.

Mudding

Mud bogging is a form of off-road motorsport popular in Canada and the United States in which the goal is to drive a vehicle through a pit of mud or a track of a set length. Winners are determined by the distance traveled through the pit. However, if several vehicles are able to travel the entire length, the time taken to traverse the pit will determine the winner.

Mud Run

Mud Runs are a popular activity involving mud. Participants run a distance of 3 miles to as long as 10 miles, while crawling through mud bogs, and battling other obstacles.

Other

- Mud is used in mud wrestling as a form of entertainment.
- Mud can be used in a dunk tank.
- Dirt biking involves biking through muddy tracks and courses.
- Baseball Rubbing Mud is used to remove the sheen from new baseballs.
- Children often like to make ‘mud pies’, throw mud at each other and play barefoot and/or coat themselves in mud.
- Albuquerque and other towns across the United States such as Gillette, Wyoming hold a yearly event in which participants play volleyball in a giant mud pit.

References

- Drainage Manual: A Guide to Integrating Plant, Soil, and Water Relationships for Drainage of Irrigated Lands, Interior Dept., Bureau of Reclamation, 1993, ISBN 0-16-061623-9.
- F. Stuart Chapin III; Pamela A. Matson; Harold A. Moon (2002). Principles of Terrestrial Ecosystem Ecology. Springer. ISBN 0387954392.
- Kötke, William H. (1993). The Final Empire: The Collapse of Civilization and the Seed of the Future. Arrow Point Press. ISBN 0963378457.
- Donahue, Miller, Shickluna (1977). Soils: an introduction to soils and plant growth (4 ed.). Inglewood Cliffs, New Jersey 07632: Prentice- Hall. pp. 115, 116. ISBN 0-13-821918-4. 142
- Rout, GR; Samantaray, S; Das, P (2001). “Aluminium toxicity in plants: a review” (PDF). Agronomie. 21 (1): 4–5. doi:10.1051/agro:2001105. Retrieved 11 June 2014.
- “Magic mud food recipe - Magic mud ingredients & cooking”. Cookadvice.com. 2 September 2012. Retrieved 26 December 2012. Soil Survey Division Staff.
- Mengel, David D., Department of Agronomy, Purdue University. “Fundamentals of Soil Cation Exchange Capacity”. Retrieved 2011-05-03.

5

Study of Soil Formation Process

The process of soil formation is scientifically known as pedogenesis. This chapter lays special emphasis on this process by explaining the minutest details of this process. It also highlights the factors affecting this process. This chapter will help readers broaden their perspective in the field of soil science.

Pedogenesis

Pedogenesis (*pedon*, meaning 'soil, earth,' and *genesis*, meaning 'origin, birth') (also termed soil development, soil evolution, soil formation, and soil genesis) is the process of soil formation as regulated by the effects of place, environment, and history. Biogeochemical processes act to both create and destroy order (anisotropy) within soils. These alterations lead to the development of layers, termed soil horizons, distinguished by differences in color, structure, texture, and chemistry. These features occur in patterns of soil type distribution, forming in response to differences in soil forming factors.

Pedogenesis is studied as a branch of pedology, the study of soil in its natural environment. Other branches of pedology are the study of soil morphology, and soil classification. The study of pedogenesis is important to understanding soil distribution patterns in current (soil geography) and past (paleopedology) geologic periods.

Overview

Soil develops through a series of changes. The starting point is weathering of freshly accumulated parent material. Primitive microbes feed on simple compounds (nutrients) released by weathering, and produce acids which contribute to weathering. They also leave behind organic residues. New soils increase in depth by a combination of weathering, and further deposition. Gradually soil is able to support higher forms of plants and animals, starting with pioneer species, and proceeding to more complex plant and animal communities.

Soils deepen with accumulation of organic matter primarily due to the activities of higher plants. Topsoil deepen through soil mixing. Soils develop layers as organic matter accumulates and leaching takes place. This development of layers is the beginning of the soil profile.

Factors of Soil Formation

Russian geologist Vasily Dokuchaev (1846-1903), commonly regarded as the father of pedology, determined in 1883 that soil formation occurs over time under the influence of climate, vegetation, topography, and parent material. He demonstrated this in 1898 using the soil forming equation:

$$\text{soil} = f(cl, o, p)\ \text{tr}$$

(where *cl* = climate, *o* = organisms, *p* = biological processes) *tr* = relative time (young, mature, old)

Clorpt

In 1941, the American soil scientist Hans Jenny published Factors of Soil Formation. His equation of state differs from the Vasily Dokuchaev equation, treating time (t) as a factor, adding topographic relief (r), and pointedly leaving the ellipsis "open" for more factors (state variables) to be added as our understanding becomes more refined:

$$\text{soil} = f(Cl, o, r, p, t, \ldots)$$

Climate

Climate regulates soil formation. Soils are more developed in areas with higher rainfall and more warmth. The rate of chemical weathering increases by 2-3 times when the temperature increases by 10 degrees Celsius. Climate also affects which organisms are present, affecting the soil chemically and physically (movement of roots).

Organisms

The organisms living in and on the soil form distinct soil types. Coniferous forests have acidic leaf litter that form soils classed as inceptisols. Mixed or deciduous forests leave a larger layer of humus, changing the elements that are either leeched or accumulated in the soil, and thereby forming soils classed as alfisols. Prairies have very high humus accumulation, creating a dark, thick A horizon characteristic of mollisols.

For example, three species of land snails in the genus *Euchondrus* in the Negev desert are noted for eating lichens growing under the surface limestone rocks and slabs (endolithic lichens). They disrupt and eat the limestone. Their grazing results in the weathering of the stones, and the subsequent formation of soil. They have a significant effect on the region: the total population of snails is estimated to process between 0.7 and 1.1 metric ton per hectare per year of limestone in the Negev desert.

Parent Material

The rock from which soil is formed is called parent material. The main types are: aeolian sediments, glacial till, glacial outwash, alluvium, lacustrine sediments and residual parent material (coral or bedrock).

Pedologists see soil formation or soil properties as a function of regional climate, biota, topography, parent material, time and other variables.

Soil Forming Processes

Soils develop from parent material by various weathering processes. Organic matter accumulation, decomposition, and humification are as critically important to soil formation as weathering. The zone of humification and weathering is termed the solum.

Soil acidification resulting from soil respiration supports chemical weathering. Plants contribute to chemical weathering through root exudates.

Soils can be enriched by deposition of sediments on floodplains and alluvial fans, and by wind-borne deposits.

Soil mixing (pedoturbation) is often an important factor in soil formation. Pedoturbation includes churning clays, cryoturbation, and bioturbation. Types of bioturbation include faunal pedoturbation (animal burrowing), floral pedoturbation (root growth, tree-uprootings), and fungal pedoturbation (mycelia growth). Pedoturbation transforms soils through destratification, mixing, and sorting, as well as creating preferential flow paths for soil gas and infiltrating water. The zone of active bioturbation is termed the soil biomantle.

Soil moisture content and water flow through the soil profile support leaching of soluble constituents, and eluviation. Eluviation is the translocation of colloid material, such as organic matter, clay and other mineral compounds. Transported constituents are deposited due to differences in soil moisture and soil chemistry, especially soil pH and redox potential. The interplay of removal and deposition results in contrasting soil horizons.

Key soil forming processes especially important to macro-scale patterns of soil formation are laterization, podzolization, calcification, salinization, and gleization.

- Laterization
- Podsolization
- Calcification
- Salinization
- Gleization

Examples

A variety of mechanisms contribute to soil formation, including siltation, erosion, overpressure and lakebed succession. A specific example of the evolution of soils in prehistoric lake beds is in the Makgadikgadi Pans of the Kalahari Desert, where change in an ancient river course led to millennia of salinity buildup and formation of calcretes and silcretes.

References

- Buol, S. W.; Hole, F. D. & McCracken, R. J. (1973). Soil Genesis and Classification (First ed.). Ames, IA: Iowa State University Press. ISBN 978-0-8138-1460-5.
- Jenny, Hans (1980), The Soil Resource - Origin and Behavior, Ecological Studies 37, New York: Springer-Verlag, ISBN 978-1461261148.
- Jenny, Hans (1994). Factors of soil formation: A System of Quantitative Pedology (PDF). New York: Dover. ISBN 0-486-68128-9. Retrieved 4 September 2014.
- Stanley W. Buol, F.D. Hole and R.W. McCracken. 1997. Soil Genesis and Classification, 4th ed. Iowa State Univ. Press, Ames ISBN 0-8138-2873-2
- Francis D. Hole and J.B. Campbell. 1985. Soil landscape analysis. Totowa Rowman & Allanheld, 214 p. ISBN 0-86598-140-X

6

An Integrated Study of Soil Morphology

This chapter primarily focuses on soil morphology and soil horizon which are important facets of soil science. The chapter aims to shed light on the unexplored aspects of soil morphology to provide a better understanding of the subject matter.

Soil Morphology

Soil morphology is the field observable attributes of the soil within the various soil horizons and the description of the kind and arrangement of the horizons. C.F. Marbut championed reliance on soil morphology instead of on theories of pedogenesis for soil classification because theories of soil genesis are both ephemeral and dynamic.

The observable attributes ordinarily described in the field include the composition, form, soil structure and organization of the soil, color of the base soil and features such as mottling, distribution of roots and pores, evidence of translocated materials such as carbonates, iron, manganese, carbon and clay, and the consistence of the soil.

The observations are typically performed on a soil profile. A profile is a vertical cut, two-dimensional, in the soil and bounds one side of a pedon. The pedon is the smallest three-dimensional unit, but not less than 1 meter square on top, that captures the lateral range of variability.

Micromorphology

While soil micromorphology begins in the field with the routine and careful use of a 10x hand lens, much more can be described by careful description of thin sections made of the soil with the aid of a petrographic polarizing light microscope. The soil can be impregnated with an epoxy resin, but more commonly with a polyester resin (crystic 17449) and sliced and ground to 0.03 millimeter thickness and examined by passing light through the thin soil plasma.

Porosity

Porosity of topsoil typically decreases as grain size increases. This is due to soil aggregate formation in finer textured surface soils when subject to soil biological processes. Aggregation involves particulate adhesion and higher resistance to compaction. Typical bulk density of sandy soil is between 1.5 and 1.7 g/cm^3. This calculates to a porosity between 0.43 and 0.36. Typical bulk density of clay soil is between 1.1 and 1.3 g/cm^3.

This calculates to a porosity between 0.58 and 0.51. This seems counterintuitive because clay soils are termed *heavy*, implying *lower* porosity. Heavy apparently refers to a gravitational moisture content effect in combination with terminology that harkens back to the relative force required to pull a tillage implement through the clayey soil at field moisture content as compared to sand.

Porosity of subsurface soil is lower than in surface soil due to compaction by gravity. Porosity of 0.20 is considered normal for unsorted gravel size material at depths below the biomantle. Porosity in finer material below the aggregating influence of pedogenesis can be expected to approximate this value.

Soil porosity is complex. Traditional models regard porosity as continuous. This fails to account for anomalous features and produces only approximate results. Furthermore it cannot help model the influence of environmental factors which affect pore geometry. A number of more complex models have been proposed, including fractals, bubble theory, cracking theory, Boolean grain process, packed sphere, and numerous other models.

Soil Composition by Laboratory Methods

Soil Texture

An experienced soil scientist can determine soil texture in the field with decent accuracy, but not all soils lend themselves to accurate field determinations of soil texture. The mineral texture can be obfuscated by high soil organic matter, iron oxides, amorphous or short-range-order aluminosilicates, and carbonates. Soil texture is the relative relations of the components sand, silt, and clay. Texture is most often reported as percentages on a mass basis. Laboratory methods employ chemical pretreatments to mediate the effects of organic matter, iron oxides, amorphous or short-range-order aluminosilicates, and carbonates.

Soil Horizon

A soil horizon is a layer generally parallel to the soil crust, whose physical characteristics differ from the layers above and beneath. Each soil type usually has three or four horizons. Horizons are defined in most cases by obvious physical features, chiefly colour and texture. These may be described both in absolute terms (particle size distribution for texture, for instance) and in terms relative to the surrounding material (i.e., "coarser" or "sandier" than the horizons above and below). The differentiation of the soil into distinct horizons is largely the result of influences, such as air, water, solar radiation and plant material, originating at the soil-atmosphere interface. Since the weathering of the soil occurs first at the surface and works its way down, the uppermost layers have been changed the most, while the deepest layers are most similar to the original parent material.

Identification and description of the horizons present at a given site is the first step in soil classification at higher levels, through the use of systems such as the USDA soil taxonomy or the Australian Soil Classification. The World Reference Base for Soil Resources lists 40 diagnostic horizons. Soil scientists often dig a large hole, called a soil pit (usually several meters deep and about a meter wide) to expose soil horizons for study.

A cross-section of soil top, revealing horizons

The vertical section exposing a set of horizons, from the ground surface to the parent rock, is termed a soil profile. Most soils, especially in temperate climates, conform to a similar general pattern of horizons, often represented as "ideal" soil in diagrams. Each main horizon is denoted by a capital letter, which may then be followed by several alphanumerical modifiers highlighting particular outstanding features of the horizon. While the general O-A-B-C-R sequence seems fairly universal, some variation exists between the classification systems in different parts of the world. In addition, the exact definition of each main horizon may differ slightly – for instance, the US system uses the thickness of a horizon as a distinguishing feature, while the Australian system does not. It should be emphasised that no one system is more correct – as artificial constructs, their utility lies in their ability to accurately describe local conditions in a consistent manner. Also, many subtropical and tropical areas have soils such as oxisols or aridisols that have very different horizons from "ideal" soil, or no horizons at all.

Main Profile

Six master soil horizons are commonly recognized and are designated using the capital letters O, A, E, B, C, and R. The following horizons are listed by their position from top to bottom within the soil profile. Not all of these layers are present in every location. For instance, P horizons only form in areas which have been waterlogged for long periods of time. Soils with a history of human interference, for instance through major earthworks or regular deep ploughing, may lack distinct horizons almost completely. When examining soils in the field, attention must be paid to the local geomorphology and the historical uses to which the land has been put in order to ensure that the appropriate names are applied to the observed horizons. The horizon not listed is the O horizon which is grass and animal/plant life.

Layers

Soil generally consists of visually and texturally distinct layers, also called profiles, which can be summarized as follows from top to bottom:

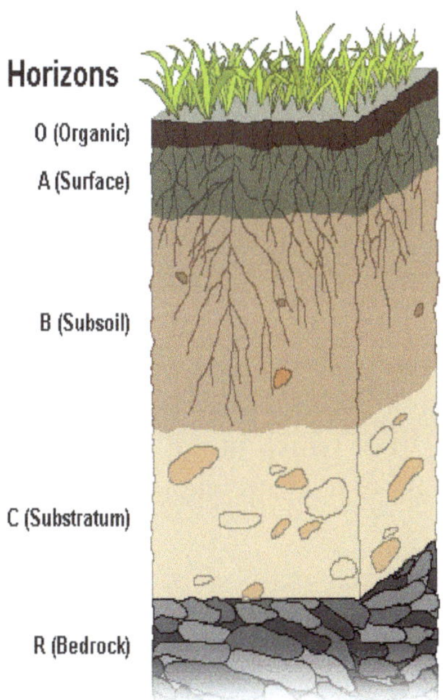

O) Organic matter: Surficial organic deposit with litter layer of plant residues in relatively non-decomposed form.

A) Surface soil: Organics mixed with mineral matter. This layer of mineral soil contains the most organic matter accumulation and soil life. This layer eluviates (is depleted of) iron, clay, aluminum, organic compounds, and other soluble constituents. When eluviation is pronounced, a lighter colored "E" subsurface soil horizon is apparent at the base of the "A" horizon. A-horizons may also be the result of a combination of soil bioturbation and surface processes that winnow fine particles from biologically mounded topsoil. In this case, the A-horizon is regarded as a "biomantle".

B) Subsoil: Subsurface layer reflecting chemical or physical alteration of parent material. This layer accumulates iron, clay, aluminum and organic compounds, a process referred to as illuviation.

C) Parent rock, also known as substratum: The parent material in sedimentary deposits. Layer of large unbroken rocks. This layer may accumulate the more soluble compounds .

R) Bedrock: The parent material in bedrock landscapes. This layer denotes the layer of partially weathered bedrock at the base of the soil profile. Unlike the above layers, R horizons largely comprise continuous masses of hard rock that cannot be excavated by hand. Soils formed *in situ* will exhibit strong similarities to this bedrock layer. These areas of bedrock are under 50 feet of the other profiles.

Soil samples illustrating horizons (subsoil on right)

O Horizon

This layer generally forms above the mineral soil or occurs in an organic soil profile. The "O" stands for organic matter. It is a surface layer dominated by the presence of large amounts of organic material derived from dead plant and/or animal residues which is in varying stages of decomposition. The O horizon is generally absent in grassland regions. The O horizon usually occurs in forested areas and is commonly referred to as the forest floor. The O horizon should be considered distinct from the layer of leaf litter covering many heavily vegetated areas, which contains no weathered mineral particles and is not part of the soil itself. O horizons may be divided into O1 and O2 categories, whereby O1 horizons contain decomposed matter whose origin can be spotted on sight (for instance, fragments of rotting leaves), and O2 horizons contain only well-decomposed organic matter, the origin of which is not readily visible. O horizons may also be divided into three subordinate O horizons denoted as: Oi, Oe, and Oa.

P Horizon

These horizons are also heavily organic, but are distinct from O horizons in that they form under waterlogged conditions. The "P" designation comes from their common name, peats. They may be divided into P1 and P2 in the same way as O Horizons. This layer accumulates iron, clay, aluminium and organic compounds, a process referred to as illuviation.

A Horizon

Surface runoff, also called nonpoint source pollution, from a farm field in Iowa during a rain storm.

The A horizon is the topmost mineral horizon, often referred to as the 'topsoil'. This layer generally contains enough partially decomposed (humified) organic matter to give the soil a color darker than that of the lower horizons. The A horizons are often coarser in texture, having lost some of the finer materials by translocation to lower horizons and by erosion. This layer is known as the zone in which the most biological activity occurs. Soil organisms such as earthworms, potworms (enchytraeids), arthropods, nematodes, fungi, and many species of bacteria and archaea are concentrated here, often in close association with plant roots. Thus the A horizon may be referred to as the biomantle. However, since biological activity extends far deeper into the soil, it cannot be used as a chief distinguishing feature of an A horizon.

E Horizon

Albeluvisol – dark surface horizon on a bleached subsurface horizon (an albic horizon) that tongues into a clay illuviation (Bt) horizon

"E", being short for eluviated, is most commonly used to label a horizon that has been significantly leached of clay, iron, and aluminum oxides, which leaves a concentration of resistant minerals, such as quartz, in the sand and silt sizes. These are present only in older, well-developed soils, and generally occur between the A and B horizons.The E horizon often has a pale color that is generally lighter in color than either the horizon above or below it. E horizons are commonly found in soils developed under forests, but are rare in soils developed under grasslands. In regions where this designation is not employed, leached layers are classified firstly as an A or B according to other characteristics, and then appended with the designation "e". In soils that contain gravels, due to animal bioturbation, a stonelayer commonly forms near or at the base of the E horizon.

The above layers may be referred to collectively as the "solum". The layers below have no collective name but are distinct in that they are noticeably less affected by surface soil-forming processes.

B Horizon

B horizons form below an O, A, or E horizon and they have undergone sufficient changes during soil genesis, such that the properties of their original parent material are no longer discernible. The B horizon is commonly referred to as the "subsoil". In humid regions, B horizons are the layers of maximum accumulation of materials such as silicate clays, iron (Fe) and aluminum (Al) oxides, and organic material. These materials typically accumulate through a process termed illuviation, wherein the materials gradually wash in from the overlying horizons. Accordingly, this layer is also referred to as the "illuviated" horizon or the "zone of accumulation". In addition, it is defined as having a distinctly different structure or consistency than the horizon(s) above and the horizon(s) below. The B horizon may also have stronger colors (higher chroma) than the A horizon. In arid and semiarid regions, calcium carbonate or calcium sulfate may accumulate in the B horizon.

As with the A horizon, the B horizon may be divided into B1, B2, and B3 types under the Australian system. B1 is a transitional horizon of the opposite nature to an A3 – dominated by the properties of the B horizons below it, but containing some A-horizon characteristics. B2 horizons have a concentration of clay, minerals, or organics and feature the highest soil development within the profile. B3 horizons are transitional between the overlying B layers and the material beneath it, whether C or D horizon.

The A3, B1, and B3 horizons are not tightly defined, and their use is generally at the discretion of the individual worker.

Plant roots penetrate through this layer, but it has little humus. It is usually brownish or reddish due to residual clay and iron oxides.

C Horizon

The C horizon (parent material) is below the B Horizon. This layer is little affected by soil-forming processes and they thus have a lack of pedological development. In other words, the C horizon is the unconsolidated material underlying the solum (A and B horizons). It may or may not be the same as the parent material from which the solum formed. The C horizon forms as the R horizon weathers and rocks break up into smaller particles. The C horizon is below the zones of greatest biological activity and it has not been sufficiently altered by soil genesis to qualify as a B horizon. In dry regions, carbonates and gypsum may be concentrated in the C horizon. While loose enough to be dug with a shovel, C horizon material often retains some of the structural features of the parent rock or geologic deposits from which it formed. The A and B layers usually originated from the C horizon. The upper layers of the C horizon may in time become a part of the solum as weathering and erosion continue. The C Horizon may contain lumps or more likely large shelves of unweathered rock, rather than being made up solely of small fragments as in the solum. It contains rocks with cracks and crevices.

Soil with broken rock fragments overlying bedrock, Sandside Bay, Caithness

D Horizon

D horizons are not universally distinguished, but in the Australian system are "any soil material below the solum that is unlike the solum in general character, is not C horizon, and cannot be given reliable designation... [it] may be recognized by the contrast in pedologic organization between it and the overlying horizons" (MacDonald et al., 1990, p. 106).

R Horizon

R horizons are the layers of partially weathered bedrock at the base of the soil profile. Unlike the above layers, R horizons are composed largely of consolidated masses of hard rock that cannot be excavated by hand. Soils formed *in situ* will exhibit strong similarities to this bedrock layer.

L Horizon

L (limnic) horizons or layers indicate mineral or organic material that has been deposited in water by precipitation or through the actions of aquatic organisms. Included are coprogenous earth (sedimentary peat), diatomaceous earth, and marl; and is usually found as a remnant of past bodies of standing water.

Subdivisions Within Master Horizons

Often, distinctive layers exist within a given master horizon, and these are indicated by a numeral following the letter designation. For example, if three different combinations of structure and colors can be seen in the B horizon, then the profile may include a sequence such as B1-B2-B3. If two different geologic parent materials (e.g., loess overlying glacial till) are present within the soil profile, the numeral 2 is placed in front of the master horizon symbols for horizons developed in the second layer of parent material. For example, a soil would have a sequence of horizons designated O-A-B-2C if C horizon developed in a different parent material. Where a layer of mineral soil material was transported by humans from a source outside the pedon, the caret symbol (^) is inserted before the master horizon designation.

Transition Horizons

Master horizon letter combinations with a slash in between, such as E/B, are used to designate transition horizons where distinct parts of the horizon have properties of E while other parts have properties of B. Transitional layers between the master horizons may be dominated by properties of one horizon but also have characteristics of another. The two master horizon letters are used to designate these transition horizons (e.g., AE, EB, BE, and BC), wherein the dominant horizon is placed before the non-dominant horizon. For example, BE is a transition to B that is more like B than E, while EB is a transition to B that is more like E than B.

Subordinate Distinctions

Subordinate horizons may occur within a master horizon and these are designated by lowercase letters following the capital master horizon letter (e.g., Ap, Bt, or Oi). Since the nature of a master horizon is only generally described by the capital master horizon letter, the lowercase letter symbols following the master horizon designation is often used to indicate more specific horizon characteristics. Subordinate distinctions include, but are not limited to, special physical characteristics and the accumulation of particular materials.

Lowercase Letter Symbols to Designate Subordinate Distinctions Within Master Horizons			
Letter	**Distinction**	**Letter**	**Distinction**
a	Organic matter, highly decomposed	b	Buried soil horizon
c	Concretions or nodules	d	Dense unconsolidated materials
e	Organic matter, intermediate decomposition	f	Frozen soil
g	Strong gleying (mottling)	h	Illuvial accumulation of organic matter
i	Organic matter, slightly decomposed	j	Jarosite (yellow sulfate mineral)
j	Cryoturbation (frost churning)	k	Accumulation of carbonates
m	Cementation or Induration	n	Accumulation of sodium
o	Accumulation of Fe and Al oxides	p	Plowing or other disturbance
q	Accumulation of silica	r	Weathered or soft bedrock
s	Illuvial organic matter and Fe and Al oxides	ss	Slickensides (shiny clay wedges)
t	Accumulation of silicate clays	u	Presence of human-manufactured materials (artifacts)
v	Plinthite (high iron, red material)	w	Distinctive color or structure without clay accumulation
x	Fragipan (high bulk density, brittle)	y	Accumulation of gypsum
z	Accumulation of soluble salts		

The US system employs largely similar suffixes, with a few important differences. For instance, "e" under the US system denotes a horizon containing "organic material of intermediate decomposition" rather than a bleached horizon. A full list of suffixes is available online as part of the USDA Soil Survey Manual.

Buried Soils

While soil formation is generally described as occurring *in situ*, as rock breaks down and is mixed with other materials, the process is often far more complicated. For instance, a fully formed profile may have developed in an area only to be buried by wind- or water-deposited sediments which later formed into another soil profile. This sort of occurrence is most common in coastal areas, and descriptions are modified by numerical prefixes. Thus, a profile containing a buried sequence could be structured O, A1, A2, B2, 2A2, 2B21, 2B22, 2C with the buried profile commencing at 2A2.

Diagnostic Surface Horizons of Mineral Soils

The diagnostic horizons at the soil surface are called epipedons ('*epi*, "over", and pedon, "soil"). The presence or absence of these particular diagnostic surface horizons help soil scientists determine the place of a soil within the Soil classification system. The epipedon includes the upper part of the soil darkened by organic matter and/or the upper eluvial horizons. It may also include the B horizon if the eluvial horizons are sufficiently darkened by organic matter. All soils must have one epipedon. Eight epipedons are recognized, but only five occur naturally over wide areas. The five naturally-occurring epipedons are called the mollic epipedon, umbric epipedon, ochric epipedon, melanic epipedon, and histic epipedon.

The mollic epipedon is a mineral surface horizon that is distinguished by its dark color, an associated accumulation of organic matter (>0.6% throughout), for its thickness (>25 cm), and for its softness even when dry. The mollic epipedon also has a high base saturation (>50%). The umbric epipedon has the same general characteristics as the mollic epipedon except it has a low base saturation (<50%). The umbric epipedon commonly develops in areas with slightly higher rainfall than mollic epipedon, and forms where the parent material has lower content of calcium and magnesium, The ochric epipedon is a mineral horizon that is either too thin, too light in color, or too low in organic matter to be considered a mollic or umbric horizon. The melanic epipedon is a mineral horizon that is very dark in color due to its high organic matter content (>0.6%). It is characteristic of soils high in such minerals as allophane, developed from volcanic ash. Melanic epipedon is extremely fluffly for a mineral soil. Histic epipedon is a 20 to 60 cm thick layer of organic soil materials (>20%) overlying a mineral soil. It is formed in wet areas, and has a layer of peat or muck with a black to dark brown color and a very low density.

Surface horizons are usually characterized by roundish granular structure that exhibits a hierarchy in which relatively large macroaggregates (0.25 to 5 mm in diameter) are composed of smaller microaggregates (2 to 250 μm). The latter, in turn, are composed of tiny packets of clay and organic matter only a few μm in size. You may easily demonstrate the existence of this hierarchy of aggregation by selecting a few of the largest aggregates in a soil and gently crumbling them into many smaller-sized pieces. You will find that even the smallest specks of soil usually are not individual particles but can be rubbed into a smear of still smaller particles of silt, clay, and humus. At each level in the hierarchy of aggregates, different factors are responsible for binding together the subunits.

Diagnostic Subsurface Horizons

Many subsurface diagnostic horizons are used to characterize different soils in soil taxonomy. Each diagnostic horizon provides a characteristic that helps place a soil in its proper class in the system. A soil may or may not have a subsurface diagnostic horizon. There are approximately 20 subsurface diagnostic horizons, of which include the argillic horizon, natric horizon, kandic horizon, oxic horizon, spodic horizon, and albic horizon. The argillic horizon (designated as Bt) is characterized by a thick accumulation of silicate clays that have moved downward from the upper horizons or have formed in place. The argillic horizon often contains clays with shiny coatings termed "argillans" or "clay skins", and accompanied by less than 15% exchangeable sodium on the colloidal complex. The natric horizon (designated as Btn) similarly has silicate clay accumulation with clay skins, but the clays are accompanied by more than 15% exchangeable sodium on the colloidal complex and by columnar or prismatic soil structural units. The natric horizon is most commonly found in arid and semiarid regions. The kandic horizon (designated as Bt, Bts), sometimes referred to as "old" clays, has an accumulation of Fe and Al oxides as well as low-activity silicate clays (e.g., kaolinite), but clay skins need not be evident. The clays are low in activity as shown by their low cation-holding capacities. The oxic horizon (designated as Bo) is a very weathered horizon with a high content of Fe/Al oxides and low-activity silicate clays. It is generally physically stable, crumbly, and not very sticky, despite its high clay content. It is found most commonly in humid tropical and subtropical regions. The spodic horizon (designated as Bh, Bhs) is an illuvial and dark-colored horizon that is characterized by the accumulation of colloidal organic matter and aluminum oxide with or without iron oxide. It is commonly found in highly leached forest soils of cool humid climates, typically on sandy-textured parent materials. The albic horizon (designated as E) is an eluvial and light-colored horizon that is characterized by a low content of clay and Fe/Al oxides.

References

- Buol, Stanley W.; Southard, Randal J.; Graham, Robert C.; McDaniel, Paul A. (2003). Soil Genesis and Classification, 5th Edition. Ames, Iowa: Iowa State Press, A Blackwell Pub. Co. p. 494. ISBN 0-8138-2873-2.
- Soil Survey Staff (1993). Soil Survey Manual. Washington D.C.: U. S. Government Printing Office. Soil Conservation Service, United States Department of Agriculture Handbook 18. Retrieved 2006-11-03.
- World reference base for soil resources. Rome: Food and Agricultural Organization of the United Nations. 1998. ISBN 92-5-104141-5.
- "Soil Horizon letter designations". Encyclopædia Britannica Online. Encyclopædia Britannica. 2008. Retrieved 2008-02-02.

7

Soil Survey: An Overview

Soil survey is one of the most significant aspect related to soil science. This chapter will unfold this topic in an easy and comprehensible manner. It will also cover some related topics like pedotransfer functions, catena (soil), etc. to enhance the knowledge of the topic.

Soil Survey

Soil survey, or soil mapping, is the process of classifying soil types and other soil properties in a given area and geo-encoding such information. It applies the principles of soil science, and draws heavily from geomorphology, theories of soil formation, physical geography, and analysis of vegetation and land use patterns. Primary data for the soil survey are acquired by field sampling and by remote sensing. Remote sensing principally uses aerial photography, but LiDAR and other digital techniques are steadily gaining in popularity. In the past, a soil scientist would take hard-copies of aerial photography, topo-sheets, and mapping keys into the field with them. Today, a growing number of soil scientists bring a ruggedized tablet computer and GPS into the field with them. The tablet may be loaded with digital aerial photos, LiDAR, topography, soil geodatabases, mapping keys, and more.

The term *soil survey* may also be used as a noun to describe the published results. In the United States, these surveys were once published in book form for individual counties by the National Cooperative Soil Survey. Today, soil surveys are no longer published in book form; they are published to the web and accessed on NRCS Web Soil Survey where a person can create a custom soil survey. This allows for rapid flow of the latest soil information to the user. In the past it could take years to publish a paper soil survey. Today it takes only moments for changes to go live to the public. Also, the most current soil survey data is made available at NRCS Soil Data Mart for high end GIS users such as professional consulting companies and universities.

The information in a soil survey can be used by farmers and ranchers to help determine whether a particular soil type is suited for crops or livestock and what type of soil management might be required. An architect or engineer might use the engineering properties of a soil to determine whether it is suitable for a certain type of construction. A homeowner may even use the information for maintaining or constructing their garden, yard, or home.

Soil Survey Components

Typical information in a published county soil survey includes the following:

- a brief overview on how to use the survey
- a general soil map for comparing the sustainability of large sections of the county
- a detailed map with specific soil series outlined and indexed
- a section on the use and management of soils
- tables describing the physical features and environment of the county
- tables containing land use suitability based on standards set by the Natural Resources Conservation Service.

Catena (Soil)

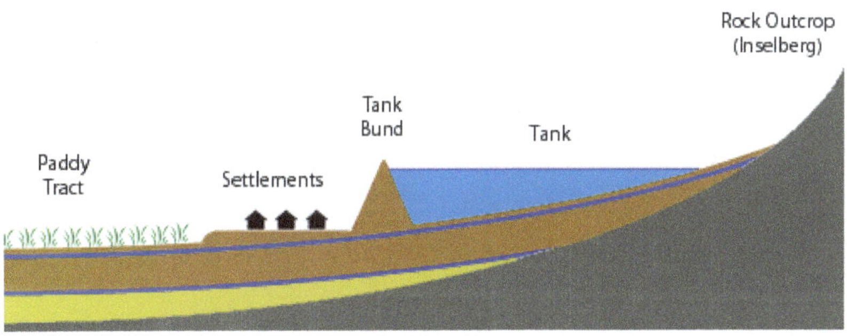

Cross Section of Dry Zone Catena of Sri Lanka showing relationship to rural land use

A Catena in soil science (pedology) is a series of distinct but co-evolving soils arrayed a slope. Each soil type or "facet" differs somewhat from its neighbours, but all occur in the same climate and on the same underlying parent material. A mature catena is in equilibrium as the processes of deposition and erosion are in balance.

The Catena Concept

The catena concept originated in Central Uganda, by chemist W.S. Martin to describe a hill slope sequence at the Bukalasa research station. The term catena (Latin: chain) was first coined by scientist Geoffrey Milne to describe these soil-topography units. The concept of a catena was developed in order to analyze the regular variation of soils across a slope. The example of this approach consists first in a structural component, the recurring pattern of certain soils in a landscape transects in which every chain element has its place in the chain, a soil has it in a landscape.

Formation

The term soil catena is used to describe the lateral variation in soils over a hillslope. A slope can be broken into sections known as a ridge, crest, midslope, and toeslope. The ridge or hilltop tends to accumulate organic matter that allows formation of an adequate thickness of soil. Steeper slope or crest sections tend to be freely drained, while at the bottom of slopes or toeslopes there is usually higher in moisture content due to poor drainage. Toeslope soils are not only higher in moisture content, but are also known to be richer in clay and organic matter.

Lithology and relief can be the primary controls on the development of certain catenas with easily disaggregated parent rock and high relief favoring particle redistribution and therefore the formation of distinct soils in particle-source and particle-deposition zones along a slope.

Catenas can also develop on low relief hillslopes, but because less potential energy is available the redistribution of mass can be dominated by subsurface flow of plasma, a combination of dissolved and suspended solids in soil water.

Open System

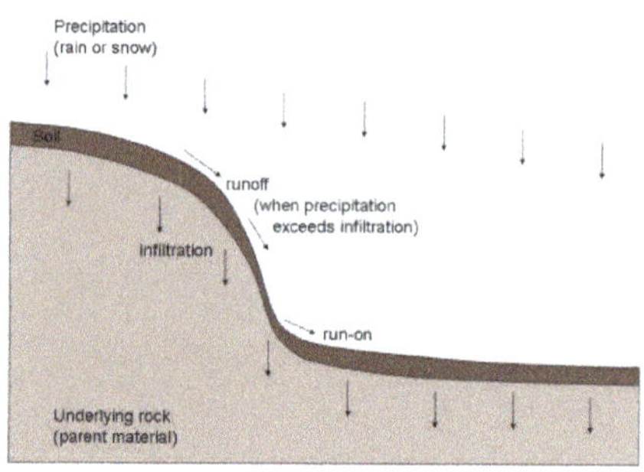

A catena is a sequence of soils down a slope, created by the balance of processes such as precipitation, infiltration and runoff.

Gullies in wet peaty soil in Scotland show where water has run off, before sinking into deeper soils at the bottom of the catena.

A catena forms when the climate, including precipitation and evaporation, is the same for the whole slope, and when sufficient time has passed for equilibrium to be reached between the processes that bring materials in to a facet and the processes that take materials away. The result is a predictable sequence of soil facets. A catena is thus an open system which has continuous input and output processes. On a steeper slope in the middle of a catena, erosion, such as by runoff, is faster, so facets are typically thinner and drier. Conversely, on a shallower slope at the top or bottom of a catena, soils are thicker and deeper. In addition, the top facets lose materials such as mineral salts when these are washed out by rain (eluviation), while the bottom facets gain materials when these are washed in (illuviation).

A catena can form on various underlying or parent materials and in different climates. On impermeable acid rocks such as metamorphic schists in a high rainfall climate like that of Western Scotland, the catena consists of thick acidic peat forming wet bog on the flatter facets, and thinner, drier, somewhat less acidic peaty podsols on the steeper facets. Thus the soil depth, acidity (pH) and soil moisture vary continuously along the slope. On a permeable basic rock such as chalk, the catena may consist of thick brown earths on the flatter facets, with thin rendzinas on the steeper slopes, while the valley bottom may include alkaline fen peat or river alluvium.

Importance

The importance of a catena is the variation of soils across a small area such as a slope. Understanding the soils that make up a catena could facilitate the mapping of soils across a given region.

Many fields of study are taken into consideration when studying catenas, studying catenas could help to understand the influence of soil hydrology on soil formation.

Catenas are found to be a great location for the study of soil science, given that the catena concept focuses on past history of the land surface, on hydrology, erosion, sediment transport, and pedogenic processes.

Liming (Soil)

Liming is the application of calcium- and magnesium-rich materials to soil in various forms, including marl, chalk, limestone, or hydrated lime. This neutralises soil acidity and increases activity of soil bacteria. However, oversupply may result in harm to plant life. Lime is a basic chemical, the effect of it makes the soil more basic thus making acidic soils neutral.

Liming of a field in Devon

The degree to which a given amount of lime per unit of soil volume will increase soil pH depends on the cation exchange capacity (CEC). Soils with low CEC will show a more marked pH increase than soils with high CEC. But the low-CEC soils will witness more rapid leaching of the added bases, and so will see a quicker return to original acidity unless additional liming is done.

Over-liming is most likely to occur on soil which has low CEC, such as sand which is deficient in buffering agents such as organic matter and clay.

Most acid soils are saturated with aluminium rather than hydrogen ions. The acidity of the soil is therefore a result of hydrolysis of aluminium. This concept of "corrected lime potential" to define the degree of base saturation in soils became the basis for procedures now used in soil testing laboratories to determine the "lime requirement" of soils.

An agricultural study at the Faculty of Forestry in Freising, Germany that compared tree stocks 2 and 20 years after liming found that liming promotes nitrate leaching and decreases the phosphorus content of some leaves.

Pedotransfer Function

In soil science, pedotransfer functions (PTF) are predictive functions of certain soil properties using data from soil surveys.

The term *pedotransfer function* was coined by Johan Bouma as *translating data we have into what we need*. The most readily available data come from soil survey, such as field morphology, soil texture, structure and pH. Pedotransfer functions add value to this basic information by translating them into estimates of other more laborious and expensively determined soil properties. These functions fill the gap between the available soil data and the properties which are more useful or required for a particular model or quality assessment. Pedotransfer functions utilize various regression analysis and data mining techniques to extract rules associating basic soil properties with more difficult to measure properties.

Although not formally recognized and named until 1989, the concept of the pedotransfer function has long been applied to estimate soil properties that are difficult to determine. Many soil science agencies have their own (unofficial) *rule of thumb* for estimating difficult-to-measure soil properties. Probably because of the particular difficulty, cost of measurement, and availability of large databases, the most comprehensive research in developing PTFs has been for the estimation of water retention curve and hydraulic conductivity.

History

The first PTF came from the study of Lyman Briggs and McLane (1907). They determined the wilting coefficient, which is defined as percentage water content of a soil when the plants growing in that soil are first reduced to a wilted condition from which they cannot recover in an approximately saturated atmosphere without the addition of water to the soil, as a function of particle-size:

$$\textit{Wilting coefficient} = 0.01\,\textit{sand} + 0.12\,\textit{silt} + 0.57\,\textit{clay}$$

With the introduction of the field capacity (FC) and permanent wilting point (PWP) concepts by Frank Veihmeyer and Arthur Hendricksen (1927), research during the period 1950-1980 attempted to correlate particle-size distribution, bulk density and organic matter content with water content at field capacity (FC), permanent wilting point (PWP), and available water capacity (AWC).

In the 1960s various papers dealt with the estimation of FC, PWP, and AWC, notably in a series of papers by Salter and Williams (1965 etc.). They explored relationships between texture classes and available water capacity, which are now known as class PTFs. They also developed functions relating the particle-size distribution to AWC, now known as continuous PTFs. They asserted that their functions could predict AWC to a mean accuracy of 16%.

In the 1970s more comprehensive research using large databases was developed. A particularly good example is the study by Hall et al. (1977) from soil in England and Wales; they established field capacity, permanent wilting point, available water content, and air capacity as a function of textural class, and as well as deriving continuous functions estimating these soil-water properties. In the USA, Gupta and Larson (1979) developed 12 functions relating particle-size distribution and organic matter content to water content at potentials ranging from -4 kPa to -1500 kPa.

With the flourishing development of models describing soil hydraulic properties and computer modelling of soil-water and solute transport, the need for hydraulic properties as inputs to these models became more evident. Clapp and Hornberger (1978) derived average values for the parameters of a power-function water retention curve, sorptivity and saturated hydraulic conductivity for different texture classes. In probably the first research of its kind, Bloemen (1977) derived empirical equations relating parameters of the Brooks and Corey hydraulic model to particle-size distribution.

Jurgen Lamp and Kneib (1981) from Germany introduced the term pedofunction, while Bouma and van Lanen (1986) used the term transfer function. To avoid confusion with the term *transfer function* used in soil physics and in many other disciplines, Johan Bouma (1989) later called it *pedotransfer function*. (A personal anecdote hinted that Arnold Bregt from Wageningen University suggested this term).

Since then, the development of hydraulic PTFs has become a boom research topic, first in the US and Europe, South America, Australia and all over the world.

Although most PTFs have been developed to predict soil hydraulic properties, they are not restricted to hydraulic properties. PTFs for estimating soil physical, mechanical, chemical and biological properties have also been developed.

Software

There are several available programs that aid determining hydraulic properties of soils using pedotransfer functions, among them are:

- SOILPAR – By Acutis and Donatelli
- ROSETTA – By Schaap et al. of the USDA, uses artificial neural networks

Soil Inference Systems

McBratney et al. (2002) introduced the concept of a soil inference system, SINFERS, where pedotransfer functions are the knowledge rules for soil inference engines. A soil inference system takes measurements with a given level of certainty (source) and by means of logically linked pedotransfer functions (organiser) infers data that is not known with minimal inaccuracy (predictor).

References

- Lindbo, David L.; Kozlowski, Deb A.; Robinson, Clay, eds. (2012). Know Soil, Know Life. Madison, Wis.: Soil Science Society of America. pp. 94–5. ISBN 9780891189541.

8

Key Branches of Soil Science

Soil science can primarily be divided into two main branches–edaphology and pedology. These branches explain the relationship of soil with its environment. Both the branches have been thoroughly discussed in this chapter to provide an in-depth understanding of soil science.

Pedology (Soil Study)

Soil Profile on Chalk at Seven Sisters Country Park, England

Pedology (*pedon*, "soil"; and *logos*, "study") is the study of soils in their natural environment. It is one of two main branches of soil science, the other being edaphology. Pedology deals with pedogenesis, soil morphology, and soil classification, while edaphology studies the way soils influence plants, fungi, and other living things.

Overview

Soil is not only a support for vegetation, but it is also the zone beneath our feet (the pedosphere) of numerous interactions between climate (water, air, temperature), soil life (micro-organisms, plants, animals) and its residues, the mineral material of the original and added rock, and its position in the landscape. During its formation and genesis, the soil profile slowly deepens and develops characteristic layers, called 'horizons', while a steady state balance is approached.

Soil users (such as agronomists) showed initially little concern in the dynamics of soil. They saw it as medium whose chemical, physical and biological properties were useful for the services of agronomic productivity . On the other hand, pedologists and geologists did not initially focus on the agronomic applications of the soil characteristics (edaphic properties) but upon its relation to the nature and history of landscapes. Today, there's an integration of the two disciplinary approaches as part of landscape and environmental sciences.

Pedologists are now also interested in the practical applications of a good understanding of pedogenesis processes (the evolution and functioning of soils), like interpreting its environmental history and predicting consequences of changes in land use, while agronomists understand that the cultivated soil is a complex medium, often resulting from several thousands of years of evolution. They understand that the current balance is fragile and that only a thorough knowledge of its history makes it possible to ensure its sustainable use.

Concepts

Important pedological concepts include:

- Complexity in soil genesis is more common than simplicity.
- Soils lie at the interface of Earth's atmosphere, biosphere, hydrosphere and lithosphere. Therefore, a thorough understanding of soils requires some knowledge of meteorology, climatology, ecology, biology, hydrology, geomorphology, geology and many other earth sciences and natural sciences.
- Contemporary soils carry imprints of pedogenic processes that were active in the past, although in many cases these imprints are difficult to observe or quantify. Thus, knowledge of paleoecology, palaeogeography, glacial geology and paleoclimatology is important for the recognition and understanding of soil genesis and constitute a basis for predicting future soil changes.
- Five major, external factors of formation (climate, organisms, relief, parent material and time), and several smaller, less identifiable ones, drive pedogenic processes and create soil patterns.
- Characteristics of soils and soil landscapes, e.g., the number, sizes, shapes and arrangements of soil bodies, each of which is characterized on the basis of soil horizons, degree of internal homogeneity, slope, aspect, landscape position, age and other properties and relationships, can be observed and measured.
- Distinctive bioclimatic regimes or combinations of pedogenic processes produce distinctive soils. Thus, distinctive, observable morphological features, e.g., illuvial clay accumulation in B horizons, are produced by certain combinations of pedogenic processes operative over varying periods of time.

- Pedogenic (soil-forming) processes act to both create and destroy order (anisotropy) within soils; these processes can proceed simultaneously. The resulting soil profile reflects the balance of these processes, present and past.
- The geological Principle of Uniformitarianism applies to soils, i.e., pedogenic processes active in soils today have been operating for long periods of time, back to the time of appearance of organisms on the land surface. These processes do, however, have varying degrees of expression and intensity over space and time.
- A succession of different soils may have developed, eroded and/or regressed at any particular site, as soil genetic factors and site factors, e.g., vegetation, sedimentation, geomorphology, change.
- There are very few old soils (in a geological sense) because they can be destroyed or buried by geological events, or modified by shifts in climate by virtue of their vulnerable position at the surface of the earth. Little of the soil continuum dates back beyond the Tertiary period and most soils and land surfaces are no older than the Pleistocene Epoch. However, preserved/lithified soils (paleosols) are an almost ubiquitous feature in terrestrial (land-based) environments throughout most of geologic time. Since they record evidence of ancient climate change, they present immense utility in understanding climate evolution throughout geologic history.
- Knowledge and understanding of the genesis of a soil is important in its classification and mapping.
- Soil classification systems cannot be based entirely on perceptions of genesis, however, because genetic processes are seldom observed and because pedogenic processes change over time.
- Knowledge of soil genesis is imperative and basic to soil use and management. Human influence on, or adjustment to, the factors and processes of soil formation can be best controlled and planned using knowledge about soil genesis.
- Soils are natural clay factories (clay includes both clay mineral structures and particles less than 2 μm in diameter). Shales worldwide are, to a considerable extent, simply soil clays that have been formed in the pedosphere and eroded and deposited in the ocean basins, to become lithified at a later date.

Famous Pedologists

- Olivier de Serres
- Vasily V. Dokuchaev
- Eugene W. Hilgard

- Hans Jenny
- Curtis F. Marbut
- Bernard Palissy

Edaphology

Edaphology is one of two main divisions of soil science, the other being pedology. Edaphology is concerned with the influence of soils on living things, particularly plants. The term is also applied to the study of how soil influences humankind's use of land for plant growth as well as man's overall use of the land. General subfields within edaphology are agricultural soil science (known by the term agrology in some regions) and environmental soil science. (Pedology deals with pedogenesis, soil morphology, and soil classification.)

In Russia, edaphology is considered equivalent to pedology, but is recognized to have an applied sense consistent with agrophysics and agrochemistry outside Russia.

History

Xenophon (431–355 BC), and Cato (234–149 BC), were early edaphologists. Xenophon noted the beneficial effect of turning a cover crop into the earth. Cato wrote De Agri Cultura ("On Farming") which recommended tillage, crop rotation and the use of legumes in the rotation to build soil nitrogen. He also devised the first soil capability classification for specific crops.

Jan Baptist van Helmont (1577–1644) performed a famous experiment, growing a willow tree in a pot of soil and supplying only rainwater for five years. The weight gained by the tree was greater than the weight loss of the soil. He concluded that the willow was made of water. Although only partly correct, his experiment reignited interest in edaphology.

Areas of Study

Agricultural Soil Science

Agricultural soil science is the application of soil chemistry, physics, and biology dealing with the production of crops. In terms of soil chemistry, it places particular emphasis on plant nutrients of importance to farming and horticulture, especially with regard to soil fertility and fertilizer components.

Physical edaphology is strongly associated with crop irrigation and drainage.

Soil husbandry is a strong tradition within agricultural soil science. Beyond preventing soil erosion and degradation in cropland, soil husbandry seeks to sustain the agricultural soil resource though the use of soil conditioners and cover crops.

Environmental Soil Science

Environmental soil science studies our interaction with the pedosphere on beyond crop production. Fundamental and applied aspects of the field address vadose zone functions, septic drain field site assessment and function, land treatment of wastewater, stormwater, erosion control, soil contamination with metals and pesticides, remediation of contaminated soils, restoration of wetlands, soil degradation, and environmental nutrient management. It also studies soil in the context of land use planning, global warming, and acid rain.

References

- Gardiner, Duane T. "Lecture 1 Chapter 1 Why Study Soils?". ENV320: Soil Science Lecture Notes. Texas A&M University-Kingsville. Retrieved 2008-01-12.
- Research Branch (1976). "Glossary of Terms in Soil Science". Publication 1459. Canada Department of Agriculture, Ottawa. Retrieved 2008-01-07.
- Whittow, John B. (1984). The Penguin Dictionary of Physical Geography. London: Penguin Books. ISBN 0-14-051094-X.
- Tseits, M. A.; B. A. Devin (2005). "Soil Science Web Resources: A Practical Guide to Search Pro-cedures and Search Engines" (PDF). Eurasian Soil Science (Translated from Pochvovedenie, no. 2, 2005, pp. 247-253) 38 (2): 223. Retrieved 2008-01-07.
- Saltini Antonio, Storia delle scienze agrarie, 4 vols, Bologna 1984-89, ISBN 88-206-2412-5, ISBN 88-206-2413-3, ISBN 88-206-2414-1, ISBN 88-206-2415-X
- Buol, Stanley W., F,D. Hole and R.W. McCracken. 1997. Soil Genesis and Classification, 4th ed. Iowa State Univ. Press, Ames ISBN 0-8138-2873-2
- Hole Francis D., J.B. Campbell. 1985. Soil landscape analysis. Totowa Rowman & Allanheld, 214 p. ISBN 0-86598-140-X
- Bockheim, J.G.; Gennadiyev, A.N.; Hammer, R.D.; Tandarich, J.P. (January 2005). "Historical development of key concepts in pedology" (PDF). Geoderma. 124 (1-2): 23–36. doi:10.1016/j.geoderma.2004.03.004. Retrieved 2 March 2016.

9

Study of Soil Mechanics and Soil Variables

This chapter provides a comprehensive insight into the mechanisms of soil variables and soil mechanics which contribute to the broadening of the existing knowledge of soil science. This chapter also incorporates related aspects like soil acidification, soil gas, soil texture, etc. and elaborates their role in soil science.

Soil Mechanics

Soil mechanics is a branch of soil physics and engineering mechanics that describes the behavior of soils. It differs from fluid mechanics and solid mechanics in the sense that soils consist of a heterogeneous mixture of fluids (usually air and water) and particles (usually clay, silt, sand, and gravel) but soil may also contain organic solids and other matter. Along with rock mechanics, soil mechanics provides the theoretical basis for analysis in geotechnical engineering, a subdiscipline of civil engineering, and engineering geology, a subdiscipline of geology. Soil mechanics is used to analyze the deformations of and flow of fluids within natural and man-made structures that are supported on or made of soil, or structures that are buried in soils. Example applications are building and bridge foundations, retaining walls, dams, and buried pipeline systems. Principles of soil mechanics are also used in related disciplines such as engineering geology, geophysical engineering, coastal engineering, agricultural engineering, hydrology and soil physics.

Slope instability issues for a temporary flood control levee in North Dakota, 2009

The Tower of Pisa – an example of a problem due to deformation of soil.

This article describes the genesis and composition of soil, the distinction between *pore water pressure* and inter-granular *effective stress*, capillary action of fluids in the soil pore spaces, *soil classification*, *seepage* and *permeability*, time dependent change of volume due to squeezing water out of tiny pore spaces, also known as *consolidation*, *shear strength* and stiffness of soils. The shear strength of soils is primarily derived from friction between the particles and interlocking, which are very sensitive to the effective stress. The article concludes with some examples of applications of the principles of soil mechanics such as slope stability, lateral earth pressure on retaining walls, and bearing capacity of foundations.

Earthwork in Germany

Fox Glacier, New Zealand: Soil produced and transported by intense weathering and erosion.

Genesis and Composition of Soils

Genesis

The primary mechanism of soil creation is the weathering of rock. All rock types (igneous rock, metamorphic rock and sedimentary rock) may be broken down into small particles to create soil. Weathering mechanisms are physical weathering, chemical weathering, and biological weathering Human activities such as excavation, blasting, and waste disposal, may also create soil. Over geologic time, deeply buried soils may be altered by pressure and temperature to become metamorphic or sedimentary rock, and if melted and solidified again, they would complete the geologic cycle by becoming igneous rock.

Physical weathering includes temperature effects, freeze and thaw of water in cracks, rain, wind, impact and other mechanisms. Chemical weatheing includes dissolution of matter composing a rock and precipitation in the form of another mineral. Clay minerals, for example can be formed by weathering of feldspar, which is the most common mineral present in igneous rock. The most common mineral constituent of silt and sand is quartz, also called silica, which has the chemical name silicon dioxide. The reason that feldspar is most common in rocks but silicon is more prevalent in soils is that feldspar is much more soluble than silica. Silt, Sand, and Gravel are basically little pieces of broken rocks.

According to the Unified Soil Classification System, silt particle sizes are in the range of 0.002 mm to 0.075 mm and sand particles have sizes in the range of 0.075 mm to 4.75 mm. Gravel particles are broken pieces of rock in the size range 4.75 mm to 100 mm. Particles larger than gravel are called cobbles and boulders.

Transport

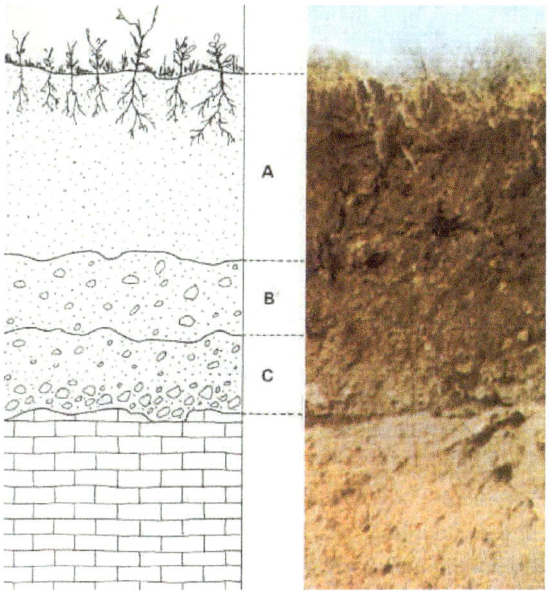

Example soil horizons. a) top soil and colluvium b) mature residual soil c) young residual soil d) weathered rock.

Soil deposits are affected by the mechanism of transport and deposition to their location. Soils that are not transported are called residual soils—they exist at the same location as the rock from which they were generated. Decomposed granite is a common example of a residual soil. The common mechanisms of transport are the actions of gravity, ice, water, and wind. Wind blown soils include dune sands and loess. Water carries particles of different size depending on the speed of the water, thus soils transported by water are graded according to their size. Silt and clay may settle out in a lake, and gravel and sand collect at the bottom of a river bed. Wind blown soil deposits (aeolian soils) also tend to be sorted according to their grain size. Erosion at the base of glaciers is powerful enough to pick up large rocks and boulders as well as soil; soils dropped by melting ice can be a well graded mixture of widely varying particle sizes. Gravity on its own may also carry particles down from the top of a mountain to make a pile of soil and boulders at the base; soil deposits transported by gravity are called colluvium.

The mechanism of transport also has a major effect on the particle shape. For example, low velocity grinding in a river bed will produce rounded particles. Freshly fractured colluvium particles often have a very angular shape.

Soil Composition

Soil Mineralogy

Silts, sands and gravels are classified by their size, and hence they may consist of a variety of minerals. Owing to the stability of quartz compared to other rock minerals, quartz is the most common constituent of sand and silt . Mica, and feldspar are other common minerals present in sands and silts. The mineral constituents of gravel may be more similar to that of the parent rock.

The common clay minerals are montmorillonite or smectite, illite, and kaolinite or kaolin. These minerals tend to form in sheet or plate like structures, with length typically ranging between 10^{-7} m and $4x10^{-6}$ m and thickness typically ranging between 10^{-9} m and $2x10^{-6}$ m, and they have a relatively large specific surface area. The specific surface area (SSA) is defined as the ratio of the surface area of particles to the mass of the particles. Clay minerals typically have specific surface areas in the range of 10 to 1,000 square meters per gram of solid. Due to the large surface area available for chemical, electrostatic, and van der Waals interaction, the mechanical behavior of clay minerals is very sensitive to the amount of pore fluid available and the type and amount of dissolved ions in the pore fluid. To anticipate the effect of clay on the way a soil will behave, it is necessary to know the kinds of clays as well as the amount present. As home builders and highway engineers know all too well, soils containing certain high-activity clays make very unstable material on which to build because they swell when wet and shrink when dry. This shrink-and-swell action can easily crack foundations and cause retaining walls to collapse. These clays also become extremely sticky and difficult to work with when they are wet. In contrast, low-activity clays, formed under different conditions, can be very stable and easy to work with.

The minerals of soils are predominantly formed by atoms of oxygen, silicon, hydrogen, and aluminum, organized in various crystalline forms. These elements along with calcium, sodium, potassium, magnesium, and carbon constitute over 99 per cent of the solid mass of soils.

Grain Size Distribution

Soils consist of a mixture of particles of different size, shape and mineralogy. Because the size of the particles obviously has a significant effect on the soil behavior, the grain size and grain size distribution are used to classify soils. The grain size distribution describes the relative proportions of particles of various sizes. The grain size is often visualized in a cumulative distribution graph which, for example, plots the percentage of particles finer than a given size as a function of size. The median grain size, D_{50}, is the size for which 50% of the particle mass consists of finer particles. Soil behavior, especially the hydraulic conductivity, tends to be dominated by the smaller particles, hence, the term “effective size”, denoted by D_{10}, , is defined as the size for which 10% of the particle mass consists of finer particles.

Sands and gravels that possess a wide range of particle sizes with a smooth distribution of particle sizes are called *well graded* soils. If the soil particles in a sample are predominantly in a relatively narrow range of sizes, the sample is *uniformly graded.* If a soil sample has distinct gaps in the gradation curve, e.g., a mixture of gravel and fine sand, with no coarse sand, the sample may be *gap graded. Uniformly graded* and *gap graded* soils are both considered to be *poorly graded.* There are many methods for measuring particle size distribution. The two traditional methods are sieve analysis and hydrometer analysis.

Sieve analysis

Sieve

The size distribution of gravel and sand particles are typically measured using sieve analysis. The formal procedure is described in ASTM D6913-04(2009). A stack of sieves with accurately dimensioned holes between a mesh of wires is used to separate the particles into size bins. A known volume of dried soil, with clods broken down to individual particles, is put into the top of a stack of sieves arranged from coarse to fine. The stack of sieves is shaken for a standard period of time so that the particles are sorted into size bins. This method works reasonably well for particles in the sand and gravel size range. Fine particles tend to stick to each other, and hence the sieving process is not an effective method. If there are a lot of fines (silt and clay) present in the soil it may be necessary to run water through the sieves to wash the coarse particles and clods through.

A variety of sieve sizes are available. The boundary between sand and silt is arbitrary. According to the Unified Soil Classification System, a #4 sieve (4 openings per inch) having 4.75mm opening size separates sand from gravel and a #200 sieve with an 0.075 mm opening separates sand from silt and clay. According to the British standard, 0.063 mm is the boundary between sand and silt, and 2 mm is the boundary between sand and gravel.

Hydrometer analysis

The classification of fine-grained soils, i.e., soils that are finer than sand, is determined primarily by their Atterberg limits, not by their grain size. If it is important to determine the grain size distribution of fine-grained soils, the hydrometer test may be performed. In the hydrometer tests, the soil particles are mixed with water and shaken to produce a dilute suspension in a glass cylinder, and then the cylinder is left to sit. A hydrometer is used to measure the density of the suspension as a function of time. Clay particles may take several hours to settle past the depth of measurement of the hydrometer. Sand particles may take less than a second. Stoke's law provides the theoretical basis to calculate the relationship between sedimentation velocity and particle size. ASTM provides the detailed procedures for performing the Hydrometer test.

Clay particles can be sufficiently small that they never settle because they are kept in suspension by Brownian motion, in which case they may be classified as colloids.

Mass-Volume Relations

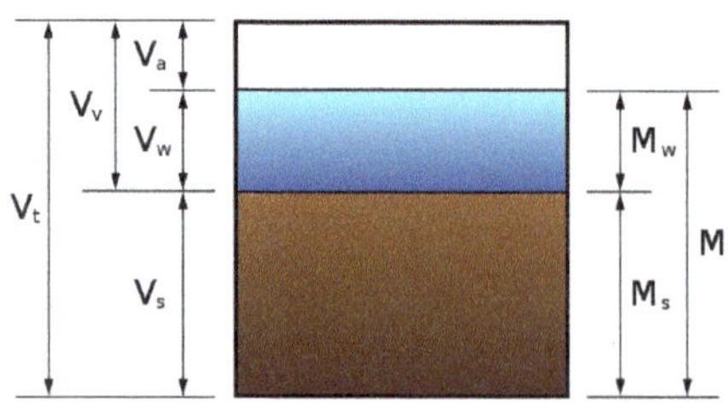

A phase diagram of soil indicating the masses and volumes of air, solid, water, and voids.

There are a variety of parameters used to describe the relative proportions of air, water and solid in a soil. This section defines these parameters and some of their interrelationships. The basic notation is as follows:

V_a, V_w, and V_s represent the volumes of air, water and solids in a soil mixture;

W_a, W_w, and W_s represent the weights of air, water and solids in a soil mixture;

M_a, M_w, and Ms represent the masses of air, water and solids in a soil mixture;

ρ_a, ρ_w, and ρ_s represent the densities of the constituents (air, water and solids) in a soil mixture;

Note that the weights, W, can be obtained by multiplying the mass, M, by the acceleration due to gravity, g; e.g., $W_s = M_s g$

Specific Gravity is the ratio of the density of one material compared to the density of pure water ($\rho_w = 1g/cm^3$).

Specific gravity of solids, $G_s = \dfrac{\rho_s}{\rho_w}$

Note that unit weights, conventionally denoted by the symbol γ may be obtained by multiplying the density (ρ) of a material by the acceleration due to gravity, g.

Density, Bulk Density, or Wet Density, ρ, are different names for the density of the mixture, i.e., the total mass of air, water, solids divided by the total volume of air water and solids (the mass of air is assumed to be zero for practical purposes):

$$\rho = \frac{M_s + M_w}{V_s + V_w + V_a} = \frac{M_t}{V_t}$$

Dry Density, ρ_d, is the mass of solids divided by the total volume of air water and solids:

$$\rho_d = \frac{M_s}{V_s + V_w + V_a} = \frac{M_s}{V_t}$$

Buoyant Density, ρ', defined as the density of the mixture minus the density of water is useful if the soil is submerged under water:

$$\rho' = \rho - \rho_w$$

where ρ_w is the density of water

Water Content, w is the ratio of mass of water to mass of solid. It is easily measured by weighing a sample of the soil, drying it out in an oven and re-weighing. Standard procedures are described by ASTM.

$$w = \frac{M_w}{M_s} = \frac{W_w}{W_s}$$

Void ratio, e, is the ratio of the volume of voids to the volume of solids:

$$e = \frac{V_V}{V_S} = \frac{V_V}{V_T - V_V} = \frac{n}{1-n}$$

Porosity, n, is the ratio of volume of voids to the total volume, and is related to the void ratio:

$$n = \frac{V_v}{V_t} = \frac{V_v}{V_s + V_v} = \frac{e}{1+e}$$

Degree of saturation, S, ,is the ratio of the volume of water to the volume of voids:

$$S = \frac{V_w}{V_v}$$

From the above definitions, some useful relationships can be derived by use of basic algebra.

$$\rho = \frac{(G_s + Se)\rho_w}{1+e}$$

$$\rho = \frac{(1+w)G_s\rho_w}{1+e}$$

$$w = \frac{Se}{G_s}$$

Effective Stress and Capillarity: Hydrostatic Conditions

Spheres immersed in water, reducing effective stress.

To understand the mechanics of soils it is necessary to understand how normal stresses and shear stresses are shared by the different phases. Neither gas nor liquid provide significant resistance to shear stress. The shear resistance of soil is provided by friction and interlocking of the particles. The friction depends on the intergranular contact stresses between solid particles. The normal stresses, on the other hand, are shared by the fluid and the particles. Although the pore air is relatively compressible, and hence takes little normal stress in most geotechnical problems, liquid water is relatively incompressible and if the voids are saturated with water, the pore water must be squeezed out in order to pack the particles closer together.

The principle of effective stress, introduced by Karl Terzaghi, states that the effective stress σ› (i.e., the average intergranular stress between solid particles) may be calculated by a simple subtraction of the pore pressure from the total stress:

$$\sigma' = \sigma - u$$

where σ is the total stress and u is the pore pressure. It is not practical to measure σ› directly, so in practice the vertical effective stress is calculated from the pore pressure and vertical total stress. The distinction between the terms pressure and stress is also important. By definition, pressure at a point is equal in all directions but stresses at a point can be different in different directions. In soil mechanics, compressive stresses and pressures are considered to be positive and tensile stresses are considered to be negative, which is different from the solid mechanics sign convention for stress.

Total Stress

For level ground conditions, the total vertical stress at a point, σ_v, on average, is the weight of everything above that point per unit area. The vertical stress beneath a uniform surface layer with density ρ, and thickness H is for example:

$$\sigma = \rho g H = \gamma H$$

where g is the acceleration due to gravity, and γ is the unit weight of the overlying layer. If there are multiple layers of soil or water above the point of interest, the vertical stress may be calculated by summing the product of the unit weight and thickness of all of the overlying layers. Total stress increases with increasing depth in proportion to the density of the overlying soil.

It is not possible to calculate the horizontal total stress in this way. Lateral earth pressures are addressed elsewhere.

Pore Water Pressure

Hydrostatic Conditions

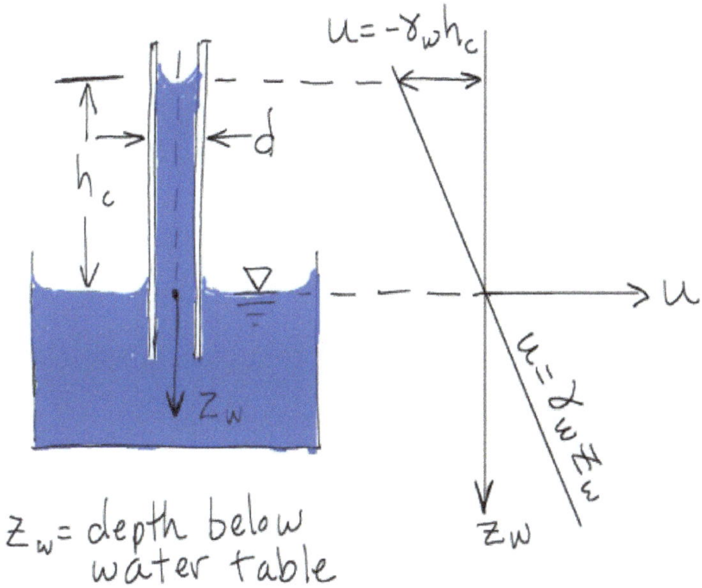

Water is drawn into a small tube by surface tension. Water pressure, u, is negative above and positive below the free water surface

If there is no pore water flow occurring in the soil, the pore water pressures will be hydrostatic. The water table is located at the depth where the water pressure is equal to the atmospheric pressure. For hydrostatic conditions, the water pressure increases linearly with depth below the water table:

$$u = \rho_w g z_w$$

where ρ_w is the density of water, and z_w is the depth below the water table.

Capillary Action

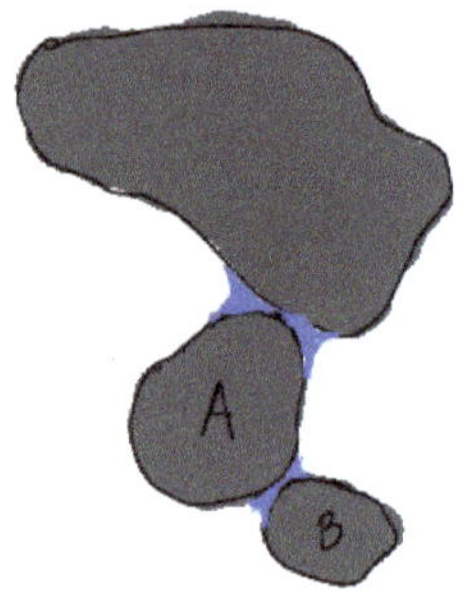

Water at particle contacts

Due to surface tension, water will rise up in a small capillary tube above a free surface of water. Likewise, water will rise up above the water table into the small pore spaces around the soil particles. In fact the soil may be completely saturated for some distance above the water table. Above the height of capillary saturation, the soil may be wet but the water content will decrease with elevation. If the water in the capillary zone is not moving, the water pressure obeys the equation of hydrostatic equilibrium, $u = \rho_w g z_w$, but note that z_w, is negative above the water table. Hence, hydrostatic water pressures are negative above the water table. The thickness of the zone of capillary saturation depends on the pore size, but typically, the heights vary between a centimeter or so for coarse sand to tens of meters for a silt or clay. In fact the pore space of soil is a uniform fractal e.g. a set of uniformly distributed D-dimensional fractals of average linear size L. For the clay soil it has been found that L=0.15 mm and D=2.7.

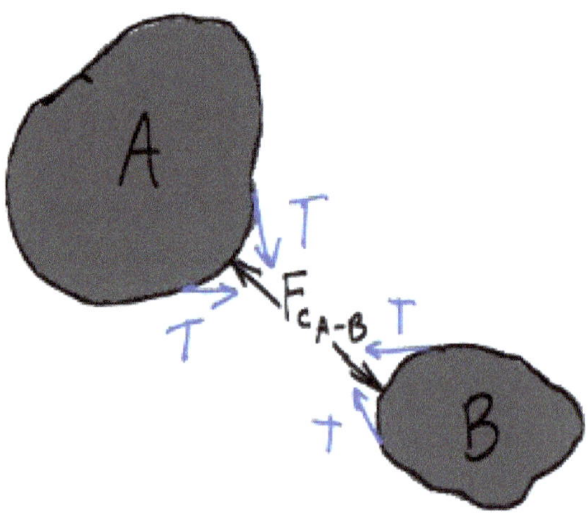

intergranular contact force due to surface tension.

The surface tension of water explains why the water does not drain out of a wet sand castle or a moist ball of clay. Negative water pressures make the water stick to the particles and pull the particles to each other, friction at the particle contacts make a sand castle stable. But as soon as a wet sand castle is submerged below a free water surface, the negative pressures are lost and the castle collapses. Considering the effective stress equation, $\sigma' = \sigma - u_{,,}$, if the water pressure is negative, the effective stress may be positive, even on a free surface (a surface where the total normal stress is zero). The negative pore pressure pulls the particles together and causes compressive particle to particle contact forces.

Negative pore pressures in clayey soil can be much more powerful than those in sand. Negative pore pressures explain why clay soils shrink when they dry and swell as they are wetted. The swelling and shrinkage can cause major distress, especially to light structures and roads.

Shrinkage caused by drying

Later sections of this article address the pore water pressures for seepage and consolidation problems.

Soil Classification

Geotechnical engineers classify the soil particle types by performing tests on disturbed (dried, passed through sieves, and remolded) samples of the soil. This provides information about the characteristics of the soil grains themselves. It should be noted that classification of the types of grains present in a soil does not account for important effects of the *structure* or *fabric* of the soil, terms that describe compactness of the particles and patterns in the arrangement of particles in a load carrying framework as well as the pore size and pore fluid distributions. Engineering geologists also classify soils based on their genesis and depositional history.

Classification of Soil Grains

In the US and other countries, the Unified Soil Classification System (USCS) is often used for soil classification. Other classification systems include the British Standard BS5390 and the AASHTO soil classification system.

Classification of Sands and Gravels

In the USCS, gravels (given the symbol *G*) and sands (given the symbol *S*) are classified according to their grain size distribution. For the USCS, gravels may be given the classification symbol *GW* (well-graded gravel), *GP* (poorly graded gravel), *GM* (gravel with a large amount of silt), or *GC* (gravel with a large amount of clay). Likewise sands may be classified as being *SW*, *SP*, *SM* or *SC*. Sands and gravels with a small but non-negligible amount of fines (5–12%) may be given a dual classification such as *SW-SC*.

Atterberg Limits

Clays and Silts, often called 'fine-grained soils', are classified according to their Atterberg limits; the most commonly used Atterberg limits are the Liquid limit (denoted by *LL* or w_l),), Plastic Limit (denoted by *PL* or w_p), and Shrinkage limit (denoted by *SL*). The shrinkage limit corresponds to a water content below which the soil will not shrink as it dries.

The liquid limit and plastic limit are arbitrary limits determined by tradition and convention. The liquid limit is determined by measuring the water content for which a groove closes after 25 blows in a standard test. Alternatively, a fall cone test apparatus may be use to measure the liquid limit. The undrained shear strength of remolded soil at the liquid limit is approximately 2 kPa. The plastic limit is the water content below which it is not possible to roll by hand the soil into 3 mm diameter cylinders. The soil cracks or breaks up as it is rolled down to this diameter. Remolded soil at the plastic limit is quite stiff, having an undrained shear strength of the order of about 200 kPa.

The Plasticity index of a particular soil specimen is defined as the difference between the Liquid limit and the Plastic limit of the specimen; it is an indicator of how much water the soil particles in the specimen can absorb. The plasticity index is the difference in water contents between states when the soil is relatively soft and the soil is relatively brittle when molded by hand.

Classification of Silts and Clays

According to the Unified Soil Classification System (USCS), silts and clays are classified by plotting the values of their plasticity index and liquid limit on a plasticity chart. The A-Line on the chart separates clays (given the USCS symbol *C*) from silts (given the symbol *M*). LL=50% separates high plasticity soils (given the modifier symbol *H*) from low plasticity soils (given the modifier symbol *L*). A soil that plots above the A-line and has LL>50% would, for example, be classified as *CH*. Other possible classifications of silts and clays are *ML, CL* and *MH*. If the Atterberg limits plot in the"hatched" region on the graph near the origin, the soils are given the dual classification 'CL-ML'.

Indices Related to Soil Strength

Liquidity Index

The effects of the water content on the strength of saturated remolded soils can be quantified by the use of the *liquidity index, LI*:

$$LI = \frac{w - PL}{LL - PL}$$

When the LI is 1, remolded soil is at the liquid limit and it has an undrained shear strength of about 2 kPa. When the soil is at the plastic limit, the LI is 0 and the undrained shear strength is about 200 kPa.

Relative Density

The density of sands (cohesionless soils) is often characterized by the relative density, D_r

$$D_r = \frac{e_{max} - e}{e_{max} - e_{min}} 100\%$$

where: e_{max} is the "maximum void ratio" corresponding to a very loose state, $_{min}$ is the "minimum void ratio" corresponding to a very dense state and e is the *in situ* void ratio. Methods used to calculate relative density are defined in ASTM D4254-00(2006).

Thus if $D_r = 100\%$ the sand or gravel is very dense, and if $D_r = 0\%$ the soil is extremely loose and unstable.

Seepage: Steady State Flow of Water

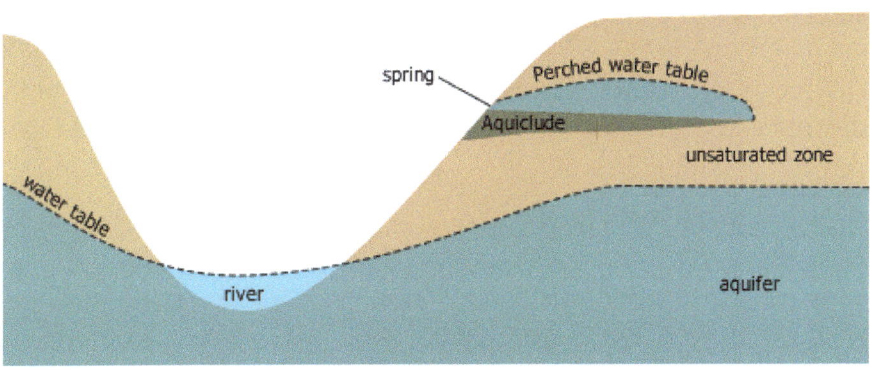

A cross section showing the water table varying with surface topography as well as a perched water table.

If fluid pressures in a soil deposit are uniformly increasing with depth according to

$u = \rho_w g z_w$ then hydrostatic conditions will prevail and the fluids will not be flowing through the soil. z_w is the depth below the water table. However, if the water table is sloping or there is a perched water table as indicated in the accompanying sketch, then seepage will occur. For steady state seepage, the seepage velocities are not varying with time. If the water tables are changing levels with time, or if the soil is in the process of consolidation, then steady state conditions do not apply.

Darcy's Law

Darcy's law states that the volume of flow of the pore fluid through a porous medium per unit time is proportional to the rate of change of excess fluid pressure with distance. The constant of proportionality includes the viscosity of the fluid and the intrinsic permeability of the soil. For the simple case of a horizontal tube filled with soil

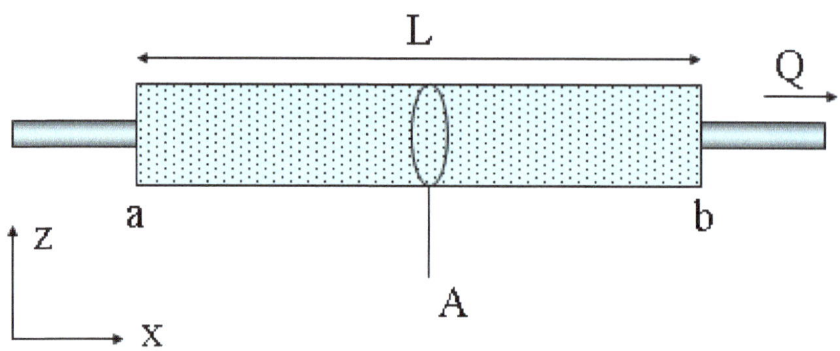

Diagram showing definitions and directions for Darcy's law.

$$Q = \frac{-KA}{\mu} \frac{(u_b - u_a)}{L}$$

The total discharge, Q (having units of volume per time, e.g., ft³/s or m³/s), is proportional to the intrinsic permeability, K, the cross sectional area, A, and rate of pore pressure change with distance, $\frac{u_b - u_a}{L}$, , and inversely proportional to the dynamic viscosity of the fluid, μ. The negative sign is needed because fluids flow from high pressure to low pressure. So if the change in pressure is negative (in the x-direction) then the flow will be positive (in the x-direction). The above equation works or a horizontal tube, but if the tube was inclined so that point b was a different elevation than point a, the equation would not work. The effect of elevation is accounted for by replacing the pore pressure by *excess pore pressure*, u_e defined as:

$$u_e = u - \rho_w g z$$

where z is the depth measured from an arbitrary elevation reference (datum). Replacing u by u_e we obtain a more general equation for flow:

$$Q = \frac{-KA}{\mu} \frac{(u_{e,b} - u_{e,a})}{L}$$

Dividing both sides of the equation by A, and expressing the rate of change of excess pore pressure as a derivative, we obtain a more general equation for the apparent velocity in the x-direction:

$$v_x = \frac{-K}{\mu} \frac{du_e}{dx}$$

where $v_x = Q / A$ has units of velocity and is called the *Darcy velocity* (or the *specific discharge, filtration velocity*, or *superficial velocity*). The *pore* or *interstitial velocity* v_{px} is the average velocity of fluid molecules in the pores; it is related to the Darcy velocity and the porosity n through the *Dupuit-Forchheimer relationship*

$$v_{px} = \frac{v_x}{n}$$

(Some authors use the term *seepage velocity* to mean the Darcy velocity, while others use it to mean the pore velocity.)

Civil engineers predominantly work on problems that involve water and predominantly work on problems on earth (in earth's gravity). For this class of problems, civil engineers will often write Darcy's law in a much simpler form:

$$v = ki$$

where k is the hydraulic conductivity, defined as $k = \frac{K\rho_w g}{\mu_w}$, and i is the hydraulic gradient. The hydraulic gradient is the rate of change of total head with distance. The total head, h at a point is defined as the height (measured relative to the datum) to which water would rise in a piezometer at that point. The total head is related to the excess water pressure by:

$$u_e = \rho_w g h + Constant$$

and the *Constant* is zero if the datum for head measurement is chosen at the same elevation as the origin for the depth, z used to calculate u_e.

Typical Values of Hydraulic Conductivity

Values of hydraulic conductivity, k , can vary by many orders of magnitude depending on the soil type. Clays may have hydraulic conductivity as small as about

$10^{-12}\frac{m}{s}$, gravels may have hydraulic conductivity up to about $10^{-1}\frac{m}{s}$.. Layering and heterogeneity and disturbance during the sampling and testing process make the accurate measurement of soil hydraulic conductivity a very difficult problem.

Flownets

Darcy's Law applies in one, two or three dimensions. In two or three dimensions, steady state seepage is described by Laplace's equation. Computer programs are available to solve this equation. But traditionally two-dimensional seepage problems were solved using a graphical procedure known as flownet. One set of lines in the flownet are in the direction of the water flow (flow lines), and the other set of lines are in the direction of constant total head (equipotential lines). Flownets may be used to estimate the quantity of seepage under dams and sheet piling.

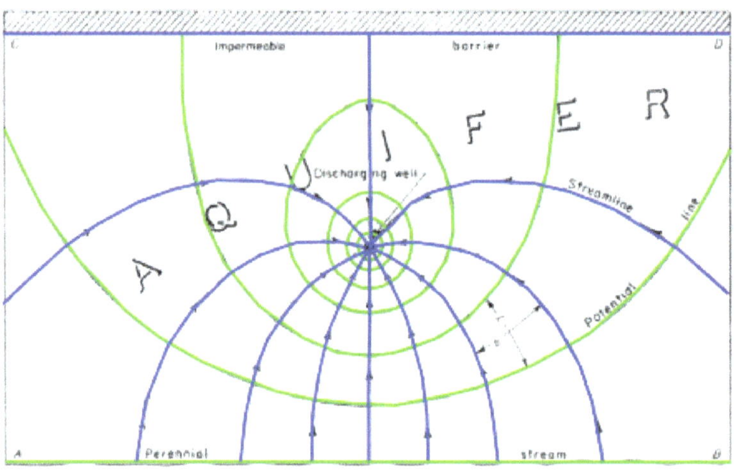

A plan flow net to estimate flow of water from a stream to a discharging well

Seepage Forces and Erosion

When the seepage velocity is great enough, erosion can occur because of the frictional drag exerted on the soil particles. Vertically upwards seepage is a source of danger on the downstream side of sheet piling and beneath the toe of a dam or levee. Erosion of the soil, known as "soil piping", can lead to failure of the structure and to sinkhole formation. Seeping water removes soil, starting from the exit point of the seepage, and erosion advances upgradient. The term "sand boil" is used to describe the appearance of the discharging end of an active soil pipe.

Seepage Pressures

Seepage in an upward direction reduces the effective stress within the soil. When the water pressure at a point in the soil is equal to the total vertical stress at that point, the effective stress is zero and the soil has no frictional resistance to deformation. For a surface layer, the vertical effective stress becomes zero within the layer when the upward hydraulic gradient is equal to the critical gradient. At zero effective stress soil has very little strength and layers of relatively impermeable soil may heave up due to the underlying water pressures. The loss in strength due to upward seepage is a common contributor to levee failures. The condition of zero effective stress associated with upward seepage is also called liquefaction, quicksand, or a boiling condition. Quicksand was so named because the soil particles move around and appear to be 'alive' (the biblical meaning of 'quick' – as opposed to 'dead'). (Note that it is not possible to be 'sucked down' into quicksand. On the contrary, you would float with about half your body out of the water.)

Consolidation: Transient Flow of Water

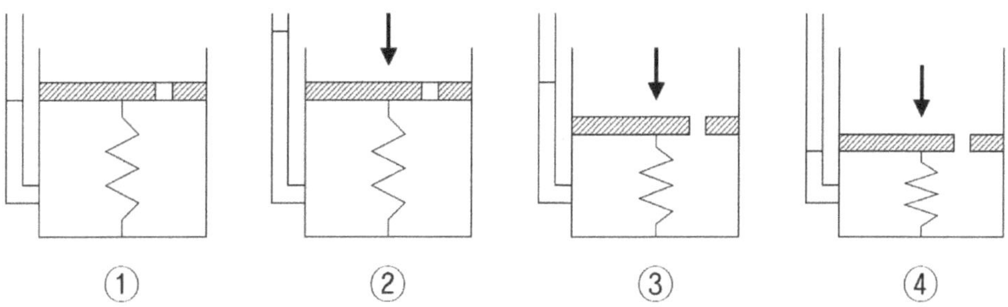

Consolidation analogy. The piston is supported by water underneath and a spring. When a load is applied to the piston, water pressure increases to support the load. As the water slowly leaks through the small hole, the load is transferred from the water pressure to the spring force.

Consolidation is a process by which soils decrease in volume. It occurs when stress is applied to a soil that causes the soil particles to pack together more tightly, therefore reducing volume. When this occurs in a soil that is saturated with water, water will be squeezed out of the soil. The time required to squeeze the water out of a thick deposit of clayey soil layer might be years. For a layer of sand, the water may be squeezed out in a matter of seconds. A building foundation or construction of a new embankment will cause the soil below to consolidate and this will cause settlement which in turn may cause distress to the building or embankment. Karl Terzaghi developed the theory of consolidation which enables prediction of the amount of settlement and the time required for the settlement to occur. Soils are tested with an oedometer test to determine their compression index and coefficient of consolidation.

When stress is removed from a consolidated soil, the soil will rebound, drawing water back into the pores and regaining some of the volume it had lost in the consolidation process. If the stress is reapplied, the soil will re-consolidate again along a recompression curve, defined by the recompression index. Soil that has been consolidated to a large pressure and has been subsequently unloaded is considered to be *overconsolidated*. The maximum past vertical effective stress is termed the *preconsolidation stress*. A soil which is currently experiencing the maximum past vertical effective stress is said to be *normally consolidated*. The *overconsolidation ratio*, (OCR) is the ratio of the maximum past vertical effective stress to the current vertical effective stress. The OCR is significant for two reasons: firstly, because the compressibility of normally consolidated soil is significantly larger than that for overconsolidated soil, and secondly, the shear behavior and dilatancy of clayey soil are related to the OCR through critical state soil mechanics; highly overconsolidated clayey soils are dilatant, while normally consolidated soils tend to be contractive.

Shear Behavior: Stiffness and Strength

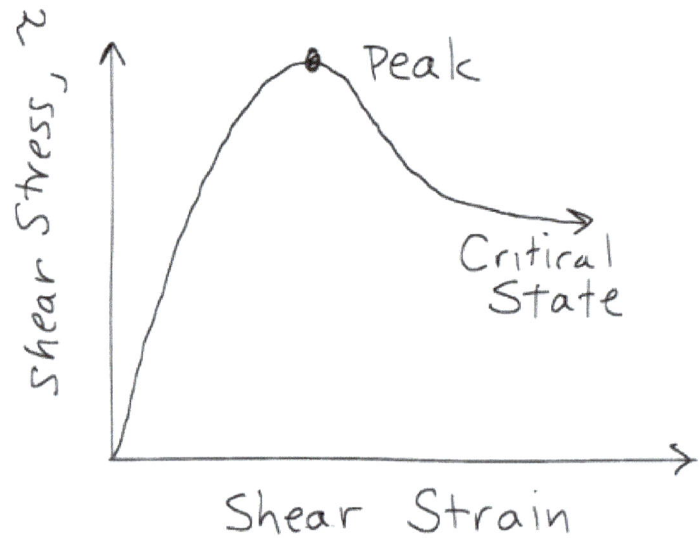

Typical stress strain curve for a drained dilatant soil

The shear strength and stiffness of soil determines whether or not soil will be stable or how much it will deform. Knowledge of the strength is necessary to determine if a slope will be stable, if a building or bridge might settle too far into the ground, and the limiting pressures on a retaining wall. It is important to distinguish between failure of a soil element and the failure of a geotechnical structure (e.g., a building foundation, slope or retaining wall); some soil elements may reach their peak strength prior to failure of the structure. Different criteria can be used to define the "shear strength" and the "yield point" for a soil element from a stress–strain curve. One may define the peak shear strength as the peak of a stress–strain curve, or the shear strength at critical state as the value after large strains when the shear resistance levels off. If the stress–strain curve does not stabilize before the end of shear strength test, the "strength" is sometimes considered to be the shear resistance at 15–20% strain. The shear strength of soil depends on many factors including the effective stress and the void ratio.

The shear stiffness is important, for example, for evaluation of the magnitude of deformations of foundations and slopes prior to failure and because it is related to the shear wave velocity. The slope of the initial, nearly linear, portion of a plot of shear stress as a function of shear strain is called the shear modulus

Friction, Interlocking and Dilation

Angle of repose

Soil is an assemblage of particles that have little to no cementation while rock (such as sandstone) may consist of an assembly of particles that are strongly cemented together by chemical bonds. The shear strength of soil is primarily due to interparticle friction and therefore, the shear resistance on a plane is approximately proportional to the effective normal stress on that plane. The angle of internal friction is thus closely related to the maximum stable slope angle, often called the angle of repose.

But in addition to friction, soil derives significant shear resistance from interlocking of grains. If the grains are densely packed, the grains tend to spread apart from each other as they are subject to shear strain. The expansion of the particle matrix due to shearing was called dilatancy by Osborne Reynolds. If one considers the energy required to shear an assembly of particles there is energy input by the shear force, T, moving a distance, x and there is also energy input by the normal force, N, as the sample expands a distance, y. Due to the extra energy required for the particles to dilate against the confining pressures, dilatant soils have a greater peak strength than contractive soils. Furthermore, as dilative soil grains dilate, they become looser (their void ratio increases), and their rate of dilation decreases until they reach a critical void ratio. Contractive soils become denser as they shear, and their rate of contraction decreases until they reach a critical void ratio.

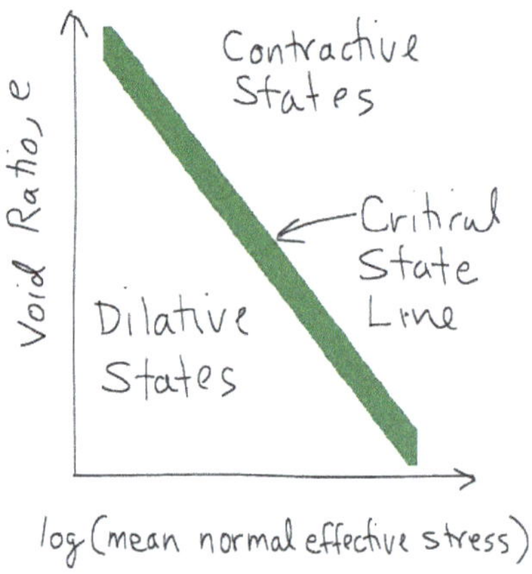

A critical state line separates the dilatant and contractive states for soil

The tendency for a soil to dilate or contract depends primarily on the confining pressure and the void ratio of the soil. The rate of dilation is high if the confining pressure is small and the void ratio is small. The rate of contraction is high if the confining pressure is large and the void ratio is large. As a first approximation, the regions of contraction and dilation are separated by the critical state line.

Failure Criteria

After a soil reaches the critical state, it is no longer contracting or dilating and the shear stress on the failure plane τ_{crit} is determined by the effective normal stress on the failure plane σ_n' and critical state friction angle ϕ_{crit}' :

$$\tau_{crit} = \sigma_n' \tan \phi_{crit}'$$

The peak strength of the soil may be greater, however, due to the interlocking (dilatancy) contribution. This may be stated:

$$\tau_{peak} = \sigma_n' \tan \phi_{peak}'$$

Where $\phi_{peak}' > \phi_{crit}'$.. However, use of a friction angle greater than the critical state value for design requires care. The peak strength will not be mobilized everywhere at the same time in a practical problem such as a foundation, slope or retaining wall. The critical state friction angle is not nearly as variable as the peak friction angle and hence it can be relied upon with confidence.

Not recognizing the significance of dilatancy, Coulomb proposed that the shear strength of soil may be expressed as a combination of adhesion and friction components:

$$\tau_f = c' + \sigma_f' \tan \phi'$$

It is now known that the c' and ϕ' parameters in the last equation are not fundamental soil properties. In particular, c' and ϕ' are different depending on the magnitude of effective stress. According to Schofield (2006), the long-standing use of c' in practice has led many engineers to wrongly believe that c' is a fundamental parameter. This assumption that c' and ϕ' are constant can lead to overestimation of peak strengths.

Structure, Fabric and Chemistry

In addition to the friction and interlocking (dilatancy) components of strength, the structure and fabric also play a significant role in the soil behavior. The structure and fabric include factors such as the spacing and arrangement of the solid particles or the amount and spatial distribution of pore water; in some cases cementitious material accumulates at particle-particle contacts. Mechanical behavior of soil is affected by the density of the particles and their structure or arrangement of the particles as well as the amount and spatial distribution of fluids present (e.g., water and air voids). Other factors include the electrical charge of the particles, chemistry of pore water, chemical bonds (i.e. cementation -particles connected through a solid substance such as recrystallized calcium carbonate)

Drained and Undrained Shear

The presence of nearly incompressible fluids such as water in the pore spaces affects the ability for the pores to dilate or contract.

If the pores are saturated with water, water must be sucked into the dilating pore spaces to fill the expanding pores (this phenomenon is visible at the beach when apparently dry spots form around feet that press into the wet sand).

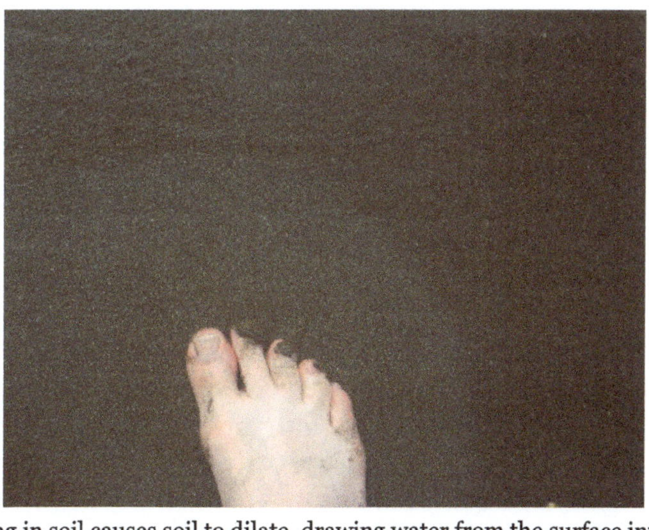

Foot pressing in soil causes soil to dilate, drawing water from the surface into the pores

Similarly, for contractive soil, water must be squeezed out of the pore spaces to allow contraction to take place.

Dilation of the voids causes negative water pressures that draw fluid into the pores, and contraction of the voids causes positive pore pressures to push the water out of the pores. If the rate of shearing is very large compared to the rate that water can be sucked into or squeezed out of the dilating or contracting pore spaces, then the shearing is called *undrained shear*, if the shearing is slow enough that the water pressures are negligible, the shearing is called *drained shear*. During undrained shear, the water pressure u changes depending on volume change tendencies. From the effective stress equation, the change in u directly effects the effective stress by the equation:

$$\sigma' = \sigma - u$$

and the strength is very sensitive to the effective stress. It follows then that the undrained shear strength of a soil may be smaller or larger than the drained shear strength depending upon whether the soil is contractive or dilative.

Shear Tests

Strength parameters can be measured in the laboratory using direct shear test, triaxial shear test, simple shear test, fall cone test and (hand) shear vane test; there are numerous other devices and variations on these devices used in practice today. Tests conducted to characterize the strength and stiffness of the soils in the ground include the Cone penetration test and the Standard penetration test.

Other Factors

The stress–strain relationship of soils, and therefore the shearing strength, is affected by:

1. *soil composition* (basic soil material): mineralogy, grain size and grain size distribution, shape of particles, pore fluid type and content, ions on grain and in pore fluid.

2. *state* (initial): Define by the initial void ratio, effective normal stress and shear stress (stress history). State can be describe by terms such as: loose, dense, overconsolidated, normally consolidated, stiff, soft, contractive, dilative, etc.

3. *structure*: Refers to the arrangement of particles within the soil mass; the manner in which the particles are packed or distributed. Features such as layers, joints, fissures, slickensides, voids, pockets, cementation, etc., are part of the structure. Structure of soils is described by terms such as: undisturbed, disturbed, remolded, compacted, cemented; flocculent, honey-combed, sin-

gle-grained; flocculated, deflocculated; stratified, layered, laminated; isotropic and anisotropic.

4. *Loading conditions*: Effective stress path -drained, undrained, and type of loading -magnitude, rate (static, dynamic), and time history (monotonic, cyclic).

Applications

Lateral Earth Pressure

Lateral earth stress theory is used to estimate the amount of stress soil can exert perpendicular to gravity. This is the stress exerted on retaining walls. A lateral earth stress coefficient, K, is defined as the ratio of lateral (horizontal) effective stress to vertical effective stress for cohesionless soils ($K=\sigma'_h/\sigma'_v$). There are three coefficients: at-rest, active, and passive. At-rest stress is the lateral stress in the ground before any disturbance takes place. The active stress state is reached when a wall moves away from the soil under the influence of lateral stress, and results from shear failure due to reduction of lateral stress. The passive stress state is reached when a wall is pushed into the soil far enough to cause shear failure within the mass due to increase of lateral stress. There are many theories for estimating lateral earth stress; some are empirically based, and some are analytically derived.

Bearing Capacity

The bearing capacity of soil is the average contact stress between a foundation and the soil which will cause shear failure in the soil. Allowable bearing stress is the bearing capacity divided by a factor of safety. Sometimes, on soft soil sites, large settlements may occur under loaded foundations without actual shear failure occurring; in such cases, the allowable bearing stress is determined with regard to the maximum allowable settlement. It is important during construction and design stage of a project to evaluate the subgrade strength. The California Bearing Ratio (CBR) test is commonly used to determine the suitability of a soil as a subgrade for design and construction. The field Plate Load Test is commonly used to predict the deformations and failure characteristics of the soil/subgrade and modulus of subgrade reaction (ks). The Modulus of subgrade reaction (ks) is used in foundation design, soil-structure interaction studies and design of highway pavements.

Slope Stability

The field of slope stability encompasses the analysis of static and dynamic stability of slopes of earth and rock-fill dams, slopes of other types of embankments, excavated slopes, and natural slopes in soil and soft rock.

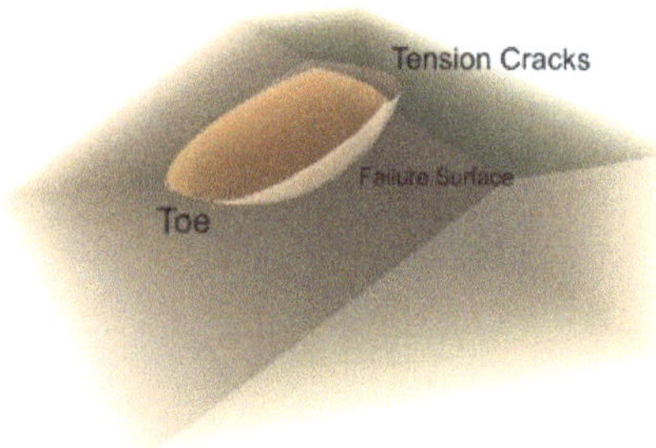

Simple slope slip section

As seen to the right, earthen slopes can develop a cut-spherical weakness zone. The probability of this happening can be calculated in advance using a simple 2-D circular analysis package... A primary difficulty with analysis is locating the most-probable slip plane for any given situation. Many landslides have been analyzed only after the fact.

Recent Developments

A recent finding in soil mechanics is that soil deformation can be described as the behavior of a dynamical system. This approach to soil mechanics is referred to as Dynamical Systems based Soil Mechanics (DSSM). DSSM holds simply that soil deformation is a Poisson process in which particles move to their final position at random shear strains.

The basis of DSSM is that soils (including sands) can be sheared till they reach a steady-state condition at which, under conditions of constant strain-rate, there is no change in shear stress, effective confining stress, and void ratio. The steady-state was formally defined by Steve J. Poulos an Associate Professor at the Soil Mechanics Department of Harvard University, who built off a hypothesis that Arthur Casagrande was formulating towards the end of his career. The steady state condition is not the same as the "critical state" condition.It differs from the critical state in that it specifies a statistically constant structure at the steady state. The steady-state values are also very slightly dependent on the strain-rate.

Many systems in nature reach steady-states and dynamical systems theory is used to describe such systems. Soil shear can also be described as a dynamical system. The physical basis of the soil shear dynamical system is a Poisson process in which particles move to the steady-state at random shear strains. Joseph generalized this—particles move to their final position (not just steady-state) at random shear-strains. Because of its origins in the steady state concept DSSM is sometimes informally called "Harvard soil mechanics."

DSSM provides for very close fits to stress–strain curves, including for sands. Because it tracks conditions on the failure plane, it also provides close fits for the post failure region of sensitive clays and silts something that other theories are not able to do. Additionally DSSM explains key relationships in soil mechanics that to date have simply been taken for granted, for example, why normalized undrained peak shear strengths vary with the log of the over consolidation ratio and why stress–strain curves normalize with the initial effective confining stress; and why in one-dimensional consolidation the void ratio must vary with the log of the effective vertical stress, why the end-of-primary curve is unique for static load increments, and why the ratio of the creep value Cα to the compression index Cc must be approximately constant for a wide range of soils.

Shear Strength (Soil)

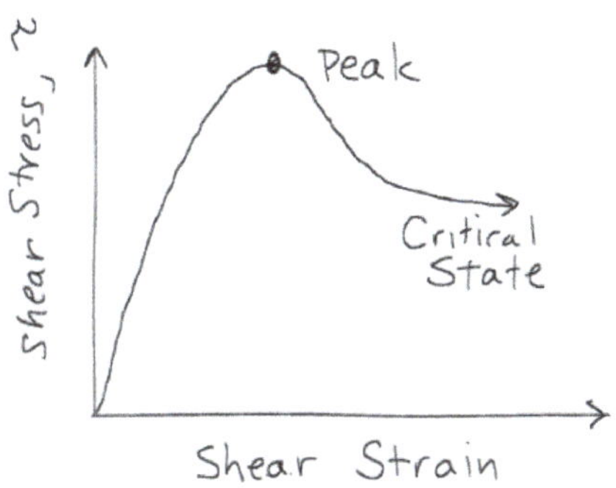

Typical stress strain curve for a drained dilatant soil

Shear strength is a term used in soil mechanics to describe the magnitude of the shear stress that a soil can sustain. The shear resistance of soil is a result of friction and interlocking of particles, and possibly cementation or bonding at particle contacts. Due to interlocking, particulate material may expand or contract in volume as it is subject to shear strains. If soil expands its volume, the density of particles will decrease and the strength will decrease; in this case, the peak strength would be followed by a reduction of shear stress. The stress-strain relationship levels off when the material stops expanding or contracting, and when interparticle bonds are broken. The theoretical state at which the shear stress and density remain constant while the shear strain increases may be called the critical state, steady state, or residual strength.

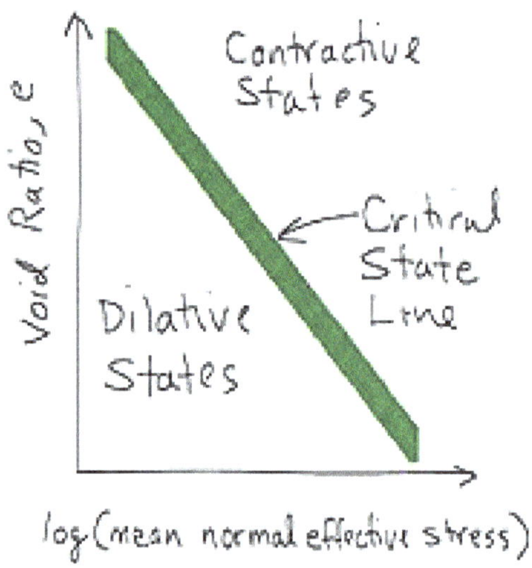

A critical state line separates the dilatant and contractive states for soil

The volume change behavior and interparticle friction depend on the density of the particles, the intergranular contact forces, and to a somewhat lesser extent, other factors such as the rate of shearing and the direction of the shear stress. The average normal intergranular contact force per unit area is called the effective stress.

If water is not allowed to flow in or out of the soil, the stress path is called an *undrained stress path*. During undrained shear, if the particles are surrounded by a nearly incompressible fluid such as water, then the density of the particles cannot change without drainage, but the water pressure and effective stress will change. On the other hand, if the fluids are allowed to freely drain out of the pores, then the pore pressures will remain constant and the test path is called a *drained stress path*. The soil is free to dilate or contract during shear if the soil is drained. In reality, soil is partially drained, somewhere between the perfectly undrained and drained idealized conditions.

The shear strength of soil depends on the effective stress, the drainage conditions, the density of the particles, the rate of strain, and the direction of the strain.

For undrained, constant volume shearing, the Tresca theory may be used to predict the shear strength, but for drained conditions, the Mohr–Coulomb theory may be used.

Two important theories of soil shear are the critical state theory and the steady state theory. There are key differences between the critical state condition and the steady state condition and the resulting theory corresponding to each of these conditions.

Factors Controlling Shear Strength of Soils

The stress-strain relationship of soils, and therefore the shearing strength, is affected (Poulos 1989) by:

1. soil composition (basic soil material): mineralogy, grain size and grain size distribution, shape of particles, pore fluid type and content, ions on grain and in pore fluid.

2. state (initial): Defined by the initial void ratio, effective normal stress and shear stress (stress history). State can be described by terms such as: loose, dense, overconsolidated, normally consolidated, stiff, soft, contractive, dilative, etc.

3. structure: Refers to the arrangement of particles within the soil mass; the manner the particles are packed or distributed. Features such as layers, joints, fissures, slickensides, voids, pockets, cementation, etc., are part of the structure. Structure of soils is described by terms such as: undisturbed, disturbed, remolded, compacted, cemented; flocculent, honey-combed, single-grained; flocculated, deflocculated; stratified, layered, laminated; isotropic and anisotropic.

4. Loading conditions: Effective stress path, i.e., drained, and undrained; and type of loading, i.e., magnitude, rate (static, dynamic), and time history (monotonic, cyclic).

Undrained Strength

This term describes a type of shear strength in soil mechanics as distinct from drained strength.

Conceptually, there is no such thing as *the* undrained strength of a soil. It depends on a number of factors, the main ones being:

- Orientation of stresses
- Stress path
- Rate of shearing
- Volume of material (like for fissured clays or rock mass)

Undrained strength is typically defined by Tresca theory, based on Mohr's circle as:

$$\sigma_1 - \sigma_3 = 2S_u$$

Where:

σ_1 is the major principal stress

σ_3 is the minor principal stress

τ is the shear strength $(\sigma_1 - \sigma_3)/2$

hence, $\tau = S_u$ (or sometimes c_u), the undrained strength.

It is commonly adopted in limit equilibrium analyses where the rate of loading is very much greater than the rate at which pore water pressures, that are generated due to the action of shearing the soil, may dissipate. An example of this is rapid loading of sands during an earthquake, or the failure of a clay slope during heavy rain, and applies to most failures that occur during construction.

As an implication of undrained condition, no elastic volumetric strains occur, and thus Poisson's ratio is assumed to remain 0.5 throughout shearing. The Tresca soil model also assumes no plastic volumetric strains occur. This is of significance in more advanced analyses such as in finite element analysis. In these advanced analysis methods, soil models other than Tresca may be used to model the undrained condition including Mohr-Coulomb and critical state soil models such as the modified Cam-clay model, provided Poisson's ratio is maintained at 0.5.

One relationship used extensively by practicing engineers is the empirical observation that the ratio of the undrained shear strength c to the original consolidation stress p' is approximately a constant for a given Over Consolidation Ratio (OCR). This relationship was first formalized by (Henkel 1960) and (Henkel & Wade 1966) who also extended it to show that stress-strain characteristics of remolded clays could also be normalized with respect to the original consolidation stress. The constant c/p relationship can also be derived from theory for both critical-state and steady-state soil mechanics (Joseph 2012). This fundamental, normalization property of the stress-strain curves is found in many clays, and was refined into the empirical SHANSEP (stress history and normalized soil engineering properties) method.(Ladd & Foott 1974).

Drained Shear Strength

The drained shear strength is the shear strength of the soil when pore fluid pressures, generated during the course of shearing the soil, are able to dissipate during shearing. It also applies where no pore water exists in the soil (the soil is dry) and hence pore fluid pressures are negligible. It is commonly approximated using the Mohr-Coulomb equation. (It was called "Coulomb's equation" by Karl von Terzaghi in 1942.) (Terzaghi 1942) combined it with the principle of effective stress.

In terms of effective stresses, the shear strength is often approximated by:

$\tau = \sigma› \tan(\varphi›) + c›$

Where $\sigma› = (\sigma - u)$, is defined as the effective stress. σ is the total stress applied normal to the shear plane, and u is the pore water pressure acting on the same plane.

$\varphi›$ = the effective stress friction angle, or the'angle of internal friction' after Coulomb friction. The coefficient of friction μ is equal to tan(φ'). Different values of friction angle can be defined, including the peak friction angle, φ'_p, the critical state friction angle, φ'_{cv}, or residual friction angle, φ'_r.

c' = is called cohesion, however, it usually arises as a consequence of forcing a straight line to fit through measured values of (τ,σ')even though the data actually falls on a curve. The intercept of the straight line on the shear stress axis is called the cohesion. It is well known that the resulting intercept depends on the range of stresses considered: it is not a fundamental soil property. The curvature (nonlinearity) of the failure envelope occurs because the dilatancy of closely packed soil particles depends on confining pressure.

Critical State Theory

A more advanced understanding of the behaviour of soil undergoing shearing lead to the development of the critical state theory of soil mechanics (Roscoe, Schofield & Wroth 1958). In critical state soil mechanics, a distinct shear strength is identified where the soil undergoing shear does so at a constant volume, also called the 'critical state'. Thus there are three commonly identified shear strengths for a soil undergoing shear:

- Peak strength τp_p
- Critical state or constant volume strength τcv_{cv}
- Residual strength τr_r

The peak strength may occur before or at critical state, depending on the initial state of the soil particles being sheared:

- A loose soil will contract in volume on shearing, and may not develop any peak strength above critical state. In this case 'peak' strength will coincide with the critical state shear strength, once the soil has ceased contracting in volume. It may be stated that such soils do not exhibit a distinct 'peak strength'.
- A dense soil may contract slightly before granular interlock prevents further contraction (granular interlock is dependent on the shape of the grains and their initial packing arrangement). In order to continue shearing once granular interlock has occurred, the soil must dilate (expand in volume). As additional shear force is required to dilate the soil, a 'peak' strength occurs. Once this peak strength caused by dilation has been overcome through continued shearing, the resistance provided by the soil to the applied shear stress reduces (termed "strain softening"). Strain softening will continue until no further changes in volume of the soil occur on continued shearing. Peak strengths are also observed in overconsolidated clays where the natural fabric of the soil must be destroyed prior to reaching constant volume shearing. Other effects that result in peak strengths include cementation and bonding of particles.

The constant volume (or critical state) shear strength is said to be extrinsic to the soil, and independent of the initial density or packing arrangement of the soil grains. In this state the grains being sheared are said to be 'tumbling' over one another, with no significant granular interlock or sliding plane development affecting the resistance to shearing. At this point, no inherited fabric or bonding of the soil grains affects the soil strength.

The residual strength occurs for some soils where the shape of the particles that make up the soil become aligned during shearing (forming a slickenside), resulting in reduced resistance to continued shearing (further strain softening). This is particularly true for most clays that comprise plate-like minerals, but is also observed in some granular soils with more elongate shaped grains. Clays that do not have plate-like minerals (like allophanic clays) do not tend to exhibit residual strengths.

Use in practice: If one is to adopt critical state theory and take c' = 0; τ_p may be used, provided the level of anticipated strains are taken into account, and the effects of potential rupture or strain softening to critical state strengths are considered. For large strain deformation, the potential to form slickensided surface with a φ'_r should be considered (such as pile driving).

The Critical State occurs at the quasi-static strain rate. It does not allow for differences in shear strength based on different strain rates. Also at the critical state, there is no particle alignment or specific soil structure.

Almost as soon as it was first introduced, the critical state concept has been subject to much criticism--chiefly its inability to match readily available test data from testing a wide variety of soils. This is primarily due to the theories inability to account for particle structure. A major consequence of this is its inability to model strain-softening post peak commonly observed in contractive soils that have anisotropic grain shapes/properties. Further, an assumption commonly made to make the model mathematically tractable is that shear stress cannot cause volumetric strain nor volumetric stress cause shear strain. Since this is not the case in reality, it is an additional cause of the poor matches to readily available empirical test data. Additionally, critical state elasto-plastic models assume that elastic strains drives volumetric changes. Since this too is not the case in real soils, this assumption results in poor fits to volume and pore pressure change data.

Steady State (Dynamical Systems Based Soil Shear)

A refinement of the critical state concept is the steady state concept.

The steady state strength is defined as the shear strength of the soil when it is at the steady state condition. The steady state condition is defined (Poulos 1981) as "that state in which the mass is continuously deforming at constant volume, constant normal effective stress, constant shear stress, and constant velocity." Steve J. Poulos, then an Associate Professor of the Soil Mechanics Department of Harvard University, built off a hypothesis that Arthur Casagrande was formulating towards the end of his career.(Poulos 1981) Steady state based soil mechanics is sometimes called "Harvard soil mechanics". The steady state condition is not the same as the "critical state" condition.

The steady state occurs only after all particle breakage if any is complete and all the particles are oriented in a statistically steady state condition and so that the shear stress needed to continue deformation at a constant velocity of deformation does not change. It applies to both the drained and the undrained case.

The steady state has a slightly different value depending on the strain rate at which it is measured. Thus the steady state shear strength at the quasi-static strain rate (the strain rate at which the critical state is defined to occur at) would seem to correspond to the critical state shear strength. However, there is an additional difference between the two states. This is that at the steady state condition the grains position themselves in the steady state structure, whereas no such structure occurs for the critical state. In the case of shearing to large strains for soils with elongated particles, this steady state structure is one where the grains are oriented (perhaps even aligned) in the direction of shear. In the case where the particles are strongly aligned in the direction of shear, the steady state corresponds to the "residual condition."

Three common misconceptions regarding the steady state are that a) it is the same as the critical state (it is not), b) that it applies only to the undrained case (it applies to all forms of drainage), and c) that it does not apply to sands (it applies to any granular material). A primer on the Steady State theory can be found in a report by Poulos (Poulos 1971). Its use in earthquake engineering is described in detail in another publication by Poulos (Poulos 1989).

The difference between the steady state and the critical state is not merely one of semantics as is sometimes thought, and it is incorrect to use the two terms/concepts interchangeably. The additional requirements of the strict definition of the steady state over and above the critical state viz. a constant deformation velocity and statistically constant structure (the steady state structure), places the steady state condition within the framework of dynamical systems theory. This strict definition of the steady state was used to describe soil shear as a dynamical system (Joseph 2009). Dynamical systems are ubiquitous in nature (the Great Red Spot on Jupiter is one example) and mathematicians have extensively studied such systems. The underlying basis of the soil shear dynamical system is simple friction (Joseph 2012).

Soil Acidification

Soil acidification is the buildup of hydrogen cations, also called protons, reducing the soil pH. This happens when a proton donor gets added to the soil. The donor can be an acid, such as nitric acid and sulfuric acid (these acids are common components of acid rain). It can also be a compound such as aluminium sulfate, which reacts in the soil to release protons. Many nitrogen compounds, which are added as fertilizer, also acidify soil over the long term because they produce nitrous and nitric acid when oxidized in the process of nitrification.

Acidification also occurs when base cations such as calcium, magnesium, potassium and sodium are leached from the soil. This leaching increases with increasing precipitation. Acid rain accelerates the leaching of bases. Plants take bases from the soil as they grow, donating a proton in exchange for each base cation. Where plant material is removed, as when a forest is logged or crops are harvested, the bases they have taken up are permanently lost from the soil.

Plant Leaves Left on Soil

Many plants produce organic acids. Where plant litter accumulates on or is incorporated to the soil, these acids (including acetic acid, humic acid, oxalic acid, and tannic acid) are liberated. This is especially acute in soils under coniferous trees such as pine, spruce and fir, which return fewer base cations to the soil than do most deciduous trees.

Rocks in the Soil

Certain parent materials also contribute to soil acidification. Granites and their allied igneous rocks are called "acidic" because they have a lot of free quartz, which produces silicic acid on weathering. Also, they have relatively low amounts of calcium and magnesium. Some sedimentary rocks such as shale and coal are rich in sulfides, which, when hydrated and oxidized, produce sulfuric acid which is much stronger than silicic acid. Many coal spoils are too acidic to support vigorous plant growth, and coal gives off strong precursors to acid rain when it is burned. Marine clays are also sulfide-rich in many cases, and such clays become very acidic if they are drained to an oxidizing state.

Pollution

Acidification may also occur from nitrogen emissions into the air, as the nitrogen may end up deposited into the soil.

Acidifying Compounds

- Aluminium sulfate
- Ammonia
- Ammonium nitrate
- Ammonium phosphate

- Ammonium sulfate
- Ferrous sulfate
- Monopotassium phosphate
- Phosphoric acid
- Urea
- Alum

Soil Texture

Soil texture is known as a qualitative classification instruments used both the field and laboratory for agricultural soils to determine classes for based on their physical texture. While classes are distinguished in the field and the class is then used to determine crop suitability and to approximate the soils responses to environmental and management conditions such as drought or calcium (lime) requirements. A qualitative rather than a quantitative tool it is a fast, simple and effective means to assess a soil's physical characteristics. The USDA soil taxonomy and WRB soil classification systems use 12 classes whereas the UK-ADAS system uses just 11.

Field Assessment Using a Hand Texture Chart

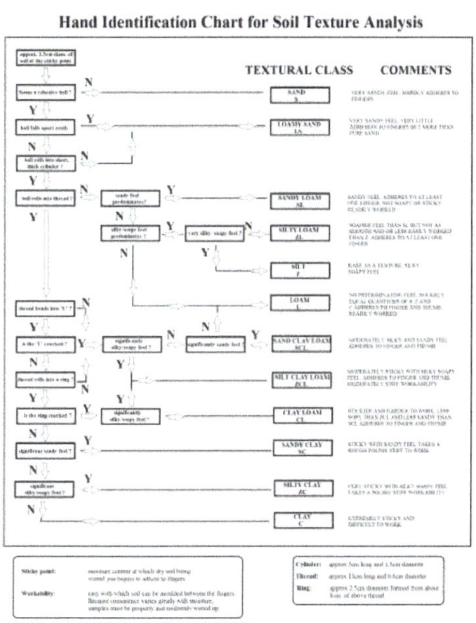

Soil texture chart

Hand analysis, while an arbitrary technique, is an extremely simple and effective means to rapidly assess and classify a soil's physical condition. Correctly executed, the procedure allows for rapid and frequent assessment of soil characteristics with little or no equipment. It is thus an extremely useful tool for identifying spatial variation both within and between plots (fields) as well as identifying progressive changes and boundaries between soil classes and orders.

The method involves taking a small sample of soil, sufficient to roll into a ball of approx 2.5 cm diameter, from just below the surface. Using a small drop of water or 'spit' the sample is then moistened to the sticky point (the point at which it begins to adhere to the finger). The ball is then molded to determine its work ability and its class according to the steps in the chart opposite.

Soil Separates

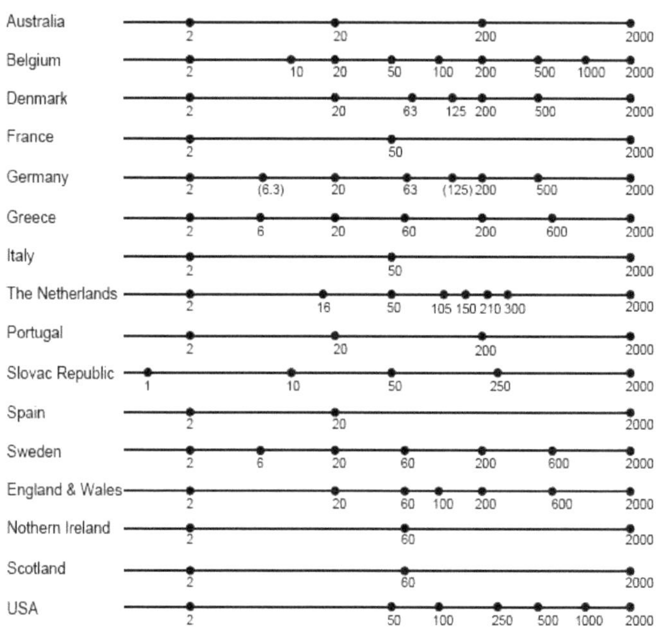

Particle size classifications used by different countries, diameters in µm

Soil separates are specific ranges of particle sizes. The smallest particles are *clay* particles and are classified as having diameters of less than 0.002 mm. The next smallest particles are *silt* particles and have diameters between 0.002 mm and 0.05 mm (in USDA soil taxonomy). The largest particles are *sand* particles and are larger than 0.05 mm in diameter. Furthermore, large sand particles can be described as *coarse*, intermediate as *medium*, and the smaller as *fine*. Other countries have their own particle size classifications.

Name of soil separate	Diameter limits (mm) (USDA classification)	Diameter limits (mm) (WRB classification)
Clay	less than 0.002	less than 0.002
Silt	0.002–0.05	0.002–0.063
Very fine sand	0.05–0.10	0.063–0.125
Fine sand	0.10–0.25	0.125–0.20
Medium sand	0.25–0.50	0.20–0.63
Coarse sand	0.50–1.00	0.63–1.25
Very coarse sand	1.00-2.00	1.25-2.00

Soil Texture Classification

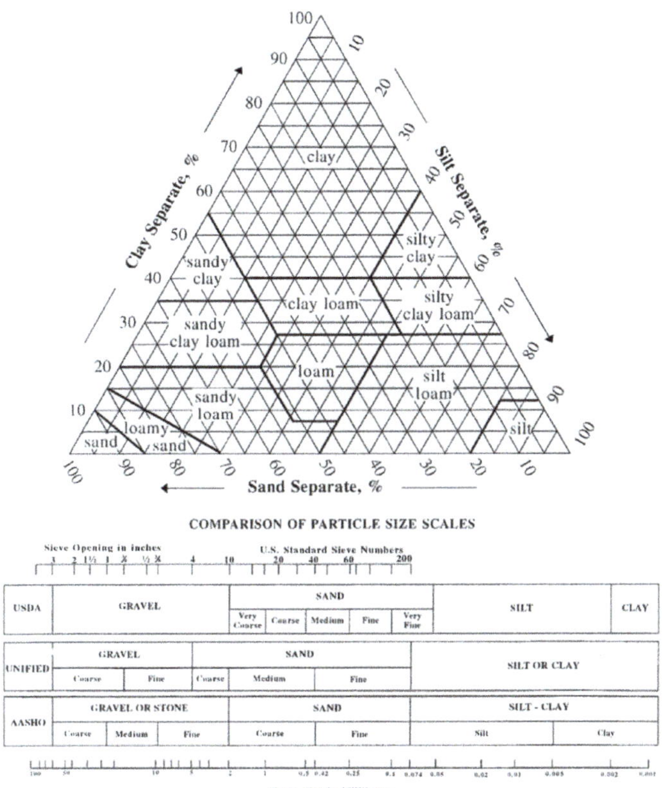

Soil texture triangle, showing the 12 major textural classes, and particle size scales as defined by the USDA.

Soil textures are classified by the fractions of each soil separate (sand, silt, and clay) present in a soil. Classifications are typically named for the primary constituent particle size or a combination of the most abundant particles sizes, e.g. "sandy clay" or "silty clay". A fourth term, loam, is used to describe a roughly equal concentration of sand, silt, and clay, and lends to the naming of even more classifications, e.g. "clay loam" or "silt loam".

In the United States, twelve major soil texture classifications are defined by the USDA.

Determining the soil textures is often aided with the use of a soil texture triangle.

History of Classification

The first classification, the International system, was first proposed by Albert Atterberg (1905), and was based on his studies in southern Sweden. Atterberg chose 20 μm for the upper limit of silt fraction because particles smaller than that size were not visible to the naked eye, the suspension could be coagulated by salts, capillary rise within 24 hours was most rapid in this fraction, and the pores between compacted particles were so small as to prevent the entry of root hairs. Commission One of the International Society of Soil Science (ISSS) recommended its use at the first International Congress of Soil Science in Washington in 1927. Australia adopted this system and according to Marshall (1947) its equal logarithmic intervals are an attractive feature worth maintaining. The USDA adopted its own system in 1938, and the FAO used the USDA system in the FAO-UNESCO world soil map and recommended its use.

Soil Resilience

Soil resilience refers to the ability of a soil to resist or recover their healthy state in response to destabilising influences - this is a subset of a notion of *environmental resilience.*

Overview

Soil resilience should first be looked at in terms of soil formation and development (pedogenesis), a continuous process taking thousands of years – this puts into context the short time that humans have so extensively utilised, changed and depended directly on soil. Pedogenesis is the result of five factors: the first two are parent material and topography, which are passive and contribute to soil mass and position; the next two are climate and the biosphere, which are active and supply the energy in soil formation. Finally, there is time.

It is the active factors in soil formation that vary so as to constitute an environmental change or shock. Over time, variations have been significant:

- Over millions of years the soil has endured varying atmospheric conditions including a complete absence of oxygen and associated behaviour of soil elements in a reducing environment, and the establishment of life - particularly of terrestrial vegetation 420 million years ago.

- Over ten thousand years and following the last ice age, though average climate has remained relatively stable, the soil has faced periods of extended wet, dry and fire.

If soil were not resilient, then in the face of past influences it would not be in any condition to support the natural and commercial services that we expect of it currently. So what do we expect of soil resilience?

- Constant or evolving state? - Do we expect that the soil to remain constant – to continue to provide the same environmental and commercial services as at present, or; that it will establish a new equilibrium?

- How long will it take? - How long do we expect soils to 'hold out' or adjust, and are we realistic about pedological time?

Australian Context

In Australia, the above questions are relevant given the strong dependence on the soil, yet the significant degradation of soils over little more than 200 years due to adoption of European styled agriculture. This is in the context of the real prospect of climate change, cyclical drought and other degrading affects.

It is in the interest of humans to sustain soils as this is the essence of our existence: the maintenance of fertile soil is "one of the most vital ecological services the living world performs"; the "mineral and organic contents of soil must be replenished constantly as plants consume soil elements and pass them up the food chain".

It is claimed by Watson (1992) that the ecosystems of Australia, which have evolved over millennia, have been decimated over the last 200 years. Our expectation has been one of ongoing environmental and commercial service, yet the practices applied have been unsustainable and have led to such soil related problems as salinity, acidity, nutrient decline, erosion and structure decline.

Barrow (1991) claims that despite decades of humans talking of an impending environmental crisis, including the breaking point of our soils' fertility, threats to the environment have continued to grow faster than the willingness to control them. Even with good intentions and best farming practices, still we are caught out by drought and wet periods, which cause unsustainable degradation.

The pressure that we impose on the soil in terms of biosphere (direct human impacts included) and climate constitute environmental change – the rate of this change compared to other changes over pedological time constitutes a shock. The resilience of the soil is limited by the rate and extent of change we impose compared with the time that soil requires to recover.

Conclusion

Soils are resilient to environmental changes and shocks – that is, they will recover from or adjust to change if sufficient 'pedological' time is allowed. The soil management practices that have been applied by humans in a short time frame are unsustainable - a declining soil health threatens human livelihood. The resilience of the soil in terms of human expectations and time frames will depend on its ability to recover to an equilibrium state once improved practices have been extensively applied.

Soil Test

Soil test may refer to one or more of a wide variety of soil analyses conducted for one of several possible reasons. Possibly the most widely conducted soil tests are those done to estimate the plant-available concentrations of plant nutrients, in order to determine fertilizer recommendations in agriculture. Other soil tests may be done for engineering (geotechnical), geochemical or ecological investigations.

Plant Nutrition

Labs, such as those at Iowa State and Colorado State Univers

In agriculture, a soil test commonly refers to the analysis of a soil sample to determine nutrient content, composition, and other characteristics such as the acidity or pH level. A soil test can determine fertility, or the expected growth potential of the soil which indicates nutrient deficiencies, potential toxicities from excessive fertility and inhibitions from the presence of non-essential trace minerals. The test is used to mimic the function of roots to assimilate minerals. The expected rate of growth is modeled by the Law of the Maximumity, recommend that a soil test contains 10-20 sample points for every 40 acres (160,000 m^2) of field. Tap water or chemicals can change the composition of the soil, and may need to be tested separately. As soil nutrients vary with depth and soil components change with time, the depth and timing of a sample may also affect results.

Composite sampling can be performed by combining soil from several locations prior to analysis. This is a common procedure, but should be used judiciously to avoid skewing results. This procedure must be done so that government sampling requirements are met. A reference map should be created to record the location and quantity of field samples in order to properly interpret test results.

Storage, Handling and Moving

Soil chemistry changes over time, as biological and chemical processes break down or combine compounds over time. These processes change once the soil is removed from its natural ecosystem (flora and fauna that penetrate the sampled area) and environment (temperature, moisture, and solar light/radiation cycles). As a result, the chemical composition analysis accuracy can be improved if the soil is analysed soon after its extraction — usually within a relative time period of 24 hours. The chemical changes in the soil can be slowed during storage and transportation by freezing it. Air drying can also preserve the soil sample for many months.

Soil Testing

Soil testing is often performed by commercial labs that offer a variety of tests, targeting groups of compounds and minerals. The advantages associated with local lab is that they are familiar with the chemistry of the soil in the area where the sample was taken. This enables technicians to recommend the tests that are most likely to reveal useful information.

Soil testing in progress

Laboratory tests often check for plant nutrients in three categories:

- Major nutrients: nitrogen (N), phosphorus (P), and potassium (K)
- Secondary nutrients: sulfur, calcium, magnesium
- Minor nutrients: iron, manganese, copper, zinc, boron, molybdenum, chlorine

Do-it-yourself kits usually only test for the three "major nutrients", and for soil acidity or pH level. Do-it-yourself kits are often sold at farming cooperatives, university labs, private labs, and some hardware and gardening stores. Electrical meters that measure pH, water content, and sometimes nutrient content of the soil are also available at many hardware stores. Laboratory tests are more accurate than tests with do-it-yourself kits and electrical meters. Here is an example soil sample report from one laboratory.

Soil testing is used to facilitate fertilizer composition and dosage selection for land employed in both agricultural and horticultural industries.

Prepaid mail-in kits for soil and ground water testing are available to facilitate the packaging and delivery of samples to a laboratory. Similarly, in 2004, laboratories began providing fertilizer recommendations along with the soil composition report.

Lab tests are more accurate, though both types are useful. In addition, lab tests frequently include professional interpretation of results and recommendations. Always refer to all proviso statements included in a lab report as they may outline any anomalies, exceptions, and shortcomings in the sampling and/or analytical process/results.

Some laboratories analyze for all 13 mineral nutrients and a dozen non-essential, potentially toxic minerals utilizing the "universal soil extractant" (ammonium bicarbonate DTPA).

Soil Contaminants

Common mineral soil contaminants include arsenic, barium, cadmium, copper, mercury, lead, and zinc.

Lead is a particularly dangerous soil component. The following table from the University of Minnesota categorizes typical soil concentration levels and their associated health risks.

Children and pregnant women should avoid contact with soil estimated total lead levels above 300 ppm		
Lead Level	**Extracted lead (ppm)**	**Estimated total lead (ppm)**
Low	<43	<500
Medium	43-126	500-1000
High	126-480	1000-3000
Very high	>480	>3000

Six gardening practices to reduce the lead risk

1. Locate gardens away from old painted structures and heavily traveled roads
2. Give planting preferences to fruiting crops (tomatoes, squash, peas, sunflowers, corn, etc.)
3. Incorporate organic materials such as finished compost, humus, and peat moss

4. Lime soil as recommended by soil test (pH 6.5 minimizes lead availability)
5. Discard old and outer leaves before eating leafy vegetables; peel root crops; wash all produce
6. Keep dust to a minimum by maintaining a mulched and/or moist soil surface

Soil Gas

Soil gases are the gases found in the air space between soil components. The primary natural soil gases include nitrogen, carbon dioxide and oxygen. The oxygen is critical because it allows for respiration of both plant roots and soil organisms. Other natural soil gases are atmospheric methane and radon. Some environmental contaminants below ground produce gas which diffuses through the soil such as from landfill wastes, mining activities, and contamination by petroleum hydrocarbons which produce volatile organic compounds. Soil gases can diffuse into buildings, the chief concerns among these pollutants are radon which is radioactive and causes cancer and methane which can be flammable at only 4.4% concentration.

Gases fill soil pores in the soil structure as water drains or is removed from a soil pore by evaporation or root absorption. The network of pores within the soil aerates, or ventilates, the soil. This aeration network becomes blocked when water enters soil pores. Not only are both soil air and soil water very dynamic parts of soil, but both are often inversely related.

Composition of air in soil and atmosphere:

- Nitrogen: Soil Air: 79.2% Atmosphere: 79.0%
- Oxygen: Soil Air: 20.6% Atmosphere: 20.9%
- Carbon Dioxide: Soil Air: 0.25% Atmosphere: 0.04%

Gas molecules in soil are in continuous thermal motion according to the kinetic theory of gases, there is also collision between molecules - a random walk.

In soil, a concentration gradient causes net movement of molecules from high concentration to low concentration, this gives the movement of gas by diffusion. Numerically, it is explained by Fick's law of diffusion.

Soil Carbon

Soil carbon includes both inorganic carbon as carbonate minerals, and as soil organic matter. Soil carbon plays a key role in the carbon cycle, and thus it is important in global climate models.

Overview

Soil carbon is present in two forms: inorganic and organic. Soil inorganic carbon consists of mineral forms of C, either from weathering of parent material, or from reaction of soil minerals with atmospheric CO_2. Carbonate minerals are the dominant form of soil carbon in desert climates. Soil organic carbon is present as soil organic matter. It includes relatively available C as fresh plant remains and relatively inert C in materials derived from plant remains: humus and charcoal.

Global Carbon Cycle

Of the 2,700 Gt of C stored in soils worldwide, 1550 GtC is organic and 950 GtC is inorganic carbon. The balance of soil carbon is held in peat and wetlands (150 GtC), and in plant litter at the soil surface (50 GtC). This compares to 780 GtC in the atmosphere, and 600 GtC in all living organisms. The oceanic pool holds 38,200 GtC.

About 60 GtC/yr is added to soil. This 60 GtC/yr is the balance of 120 GtC/yr taken out of the atmosphere by terrestrial plant photosynthesis reduced by 60 GtC/yr plant respiration. An equivalent 60 GtC/yr is respired from soil, joining the 60G tC/yr plant respiration to return to the atmosphere.

Organic Carbon

Soil organic carbon is divided between living soil biota and dead biotic material derived from biomass. Together these comprise the soil food web, with the living component sustained by the biotic material component. Soil biota includes earthworms, nematodes, protozoa, fungi, bacteria and different arthropods.

Detritus resulting from plant senescence is the major source of soil organic carbon. Plant materials, with cell walls high in cellulose and lignin, are decomposed and the not-respired carbon is retained as humus. Cellulose and starches are easily degraded, resulting in relatively short residence times. More persistent forms of organic C include lignin, humus, organic matter encapsulated in soil aggregates, and charcoal. These resist alteration and have long residence times.

Soil organic carbon tends to be concentrated in the topsoil. Topsoil ranges from 0.5% to 3.0% organic C for most upland soils. Soils with less than 0.5% organic C are mostly limited to desert areas. Soils containing greater than 12 - 18% organic C are generally classified as organic soils. High levels of organic C develop in soils supporting wetland ecology, flood deposition, fire ecology, and human activity.

Fire derived forms of carbon are present in most soils as unweathered charcoal and weathered black carbon. Soil organic C is typically 5 - 50% derived from char, with levels above 50% encountered in mollisol, chernozem, and terra preta soils.

Root exudates are another source of soil carbon. 5 - 20% of the total plant carbon fixed during photosynthesis is supplied as root exudates in support of rhizospheric mutualistic biota. Microbial populations are typically higher in the rhizosphere than in adjacent bulk soil.

Soil Health

A portable soil respiration system measuring soil CO_2 flux

Organic carbon is vital to soil capacity to provide edaphic ecosystem services. The condition of this capacity is termed soil health, a term that communicates the value of understanding soil as a living system as opposed to an abiotic component. Specific carbon related benchmarks used to evaluate soil health include CO_2 release, humus levels, and microbial metabolic activity.

Losses

The exchange of carbon between soils and the atmosphere is a significant part of the world carbon cycle. Carbon, as it relates to the organic matter of soils, is a major component of soil and catchment health. Several factors affect the variation that exists in soil organic matter and soil carbon; the most significant has, in contemporary times, been the influence of humans and agricultural systems.

Although exact quantities are difficult to measure, human activities have caused massive losses of soil organic carbon. First was the use of fire, which removes soil cover and leads to immediate and continuing losses of soil organic carbon. Tillage and drainage both expose soil organic matter to oxygen and oxidation. In the Netherlands, East Anglia, Florida, and the California Delta, subsidence of peat lands from oxidation has been severe as a result of tillage and drainage. Grazing management that exposes soil (through either excessive or insufficient recovery periods) can also cause losses of soil organic.

Managing Soil Carbon

Natural variations in SOM occur as a result of climate, organisms, parent material, time, and relief. The greatest contemporary influence has been that of humans; for example, historical SOM in Australian agricultural soils may have been twice the present range that is typically from 1.6 to 4.6 per cent.

It has long been encouraged that farmers adjust practices to maintain or increase the organic component in the soil. On one hand, practices that hasten oxidation of carbon (such as burning crop stubbles or over-cultivation) are discouraged; on the other hand, incorporation of organic material (such as in manuring) has been encouraged. Increasing soil carbon is not a straightforward matter; it is made complex by the relative activity of soil biota, which can consume and release carbon and are made more active by the addition of nitrogen fertilizers.

Data Available on soil Organic Carbon

Europe

The most homogeneous and comprehensive data on the organic carbon/matter content of European soils remain those that can be extracted and/or derived from the European Soil Database in combination with associated databases on land cover, climate, and topography. The modelled data refer to carbon content (%) in the surface horizon of soils in Europe. In an inventory on available national datasets, seven member states of the European Union have available datasets on organic carbon. In the article "Estimating soil organic carbon in Europe based on data collected through an European network" (Ecological Indicators 24, pp. 439–450), a comparison of national data with modelled data is performed. Finally, a new proposed model for estimation of soil organic carbon in agricultural soils has estimated current top SOC stock of 17.63 Gt in EU agricultural soils.

Managing for Catchment Health

Much of the contemporary literature on soil carbon relates to its role, or potential, as an atmospheric carbon sink to offset climate change. Despite this emphasis, a much wider range of soil and catchment health aspects are improved as soil carbon is increased. These benefits are difficult to quantify, due to the complexity of natural resource systems and the interpretation of what constitutes soil health; nonetheless, several benefits are proposed in the following points:

- Reduced erosion, sedimentation: increased soil aggregate stability means greater resistance to erosion; mass movement is less likely when soils are able to retain structural strength under greater moisture levels.
- Greater productivity: healthier and more productive soils can contribute to positive socio-economic circumstances.

- Cleaner waterways, nutrients and turbidity: nutrients and sediment tend to be retained by the soil rather than leach or wash off, and are so kept from waterways.
- Water balance: greater soil water holding capacity reduces overland flow and recharge to groundwater; the water saved and held by the soil remains available for use by plants.
- Climate change: Soils have the ability to retain carbon that may otherwise exist as atmospheric CO_2 and contribute to global warming.
- Greater biodiversity: soil organic matter contributes to the health of soil flora and, accordingly, the natural links with biodiversity in the greater biosphere.

Forest Soils

Forest soils constitute a large pool of carbon. Anthropogenic activities such as deforestation cause releases of carbon from this pool, which may significantly increase the concentration of greenhouse gas (GHG) in the atmosphere. Under the United Nations Framework Convention on Climate Change (UNFCCC), countries must estimate and report GHG emissions and removals, including changes in carbon stocks in all five pools (above- and below-ground biomass, dead wood, litter, and soil carbon) and associated emissions and removals from land use, land-use change and forestry activities, according to the Intergovernmental Panel on Climate Change's good practice guidance. Tropical deforestation represents nearly 25 percent of total anthropogenic GHG emissions worldwide. Deforestation, forest degradation, and changes in land management practices can cause releases of carbon from soil to the atmosphere. For these reasons, reliable estimates of soil organic carbon stock and stock changes are needed for Reducing Emissions from Deforestation and Forest Degradation and GHG reporting under the UNFCCC.

The government of Tanzania—together with the Food and Agriculture Organization of the United Nations and the financial support of the government of Finland—have implemented a forest soil carbon monitoring program to estimate soil carbon stock, using both survey and modelling-based methods.

References

- Mitchell, J.K., and Soga, K. (2005) Fundamentals of soil behavior, Third edition, John Wiley and Sons, Inc., ISBN 978-0-471-46302-3.
- Santamarina, J.C., Klein, K.A., & Fam, M.A. (2001). Soils and Waves: Particulate Materials Behavior, Characterization and Process Monitoring. Wiley. ISBN 978-0-471-49058-6.
- Terzaghi, K., Peck, R.B., Mesri, G. (1996) Soil mechanics in Engineering Practice, Third Edition, John Wiley & Sons, Inc.,ISBN 0-471-08658-4

- Bird, M. (2015). “Test procedures for biochar in soil”. In Lehmann, J.; Joseph, S. Biochar for Environmental Management (2 ed.). p. 679. ISBN 978-0-415-70415-1.
- Ruddiman, William (2007). Plows, Plagues, and Petroleum: How Humans Took Control of Climate. Princeton, NJ: Princeton University Press. ISBN 978-0-691-14634-8.
- Charman, P.E.V.; Murphy, B.W. (2000). Soils, their properties and management (2nd edn ed.). Melbourne: Oxford University Press. ISBN 978-0-19-551762-0.
- Young, A.; Young, R. (2001). Soils in the Australian landscape. Melbourne: Oxford University Press. ISBN 978-0-19-551550-3.
- Pan Y., Birdsey R., Fang J., Houghton R., Kauppi P., Kurz W., Phillips O., Shvidenko A., et al. (2011). “A Large and Persistent Carbon Sink in the World’s Forests”. Science. 333 (6045): 1201609. doi:10.1126/science.1201609.
- Lal, R. (February 2007). “Carbon Management in Agricultural Soils”. Mitigation and Adaption Strategies for Global Change. Springer. 12: 303–322. doi:10.1007/s11027-006-9036-7. Retrieved 16 January 2016.
- Lal, Rattan (2008). “Sequestration of atmospheric CO2 in global carbon pools”. Energy and Environmental Science. 1 (1): 86–100. doi:10.1039/b809492f. Retrieved 16 January 2016.
- Joseph, Paul G. (2014). “Generalized soil deformation model based on dynamical systems theory”. Geotechnical Research. 1 (1): 32–42. doi:10.1680/geores.14.00004.
- Joseph, P. G. (2012), “Physical Basis and Validation of a Constitutive Model for Soil Shear Derived from Micro-Structural Changes” (PDF), International Journal of Geomechanics, doi:10.1061/(ASCE)GM.1943-5622.0000209

10

Understanding Soil Ecology

One of the primary fields of soil science is soil ecology. This chapter studies in detail the processes and procedures of soil ecology and soil biodiversity providing a detailed account on each of these aspects related to the specified area of study.

Soil Ecology

Soil ecology is the study of the interactions among soil organisms, and between biotic and abiotic aspects of the soil environment. It is particularly concerned with the cycling of nutrients, formation and stabilization of the pore structure, the spread and vitality of pathogens, and the biodiversity of this rich biological community.

Overview

Soil is made up of a multitude of physical, chemical, and biological entities, with many interactions occurring among them. Soil is a variable mixture of broken and weathered minerals and decaying organic matter. Together with the proper amounts of air and water, it supplies, in part, sustenance for plants as well as mechanical support.

The diversity and abundance of soil life exceeds that of any other ecosystem. Plant establishment, competitiveness, and growth is governed largely by the ecology below-ground, so understanding this system is an essential component of plant sciences and terrestrial ecology.

Features of the Ecosystem

- Moisture is a major limiting factor on land. Terrestrial organisms are constantly confronted with the problem of dehydration. Transpiration or evaporation of water from plant surfaces is an energy dissipating process unique to the terrestrial environment.
- Temperature variations and extremes are more pronounced in the air than in the water medium.
- On the other hand, the rapid circulation of air throughout the globe results in a ready mixing and remarkably constant content of oxygen and carbon dioxide.
- Although soil offers solid support, air does not. Storing skeletons have been

evolved in both land plants and animals and also special means of locomotion have been evolved in the latter.

- Land, unlike the ocean, is not continuous; there are important geographical barriers to free movement.
- The nature of the substrate, although important in water is especially vital in terrestrial environment. Soil, not air, is the source of highly variable nutrients; it is a highly developed ecological subsystem.

Soil Food Web

An incredible diversity of organisms make up the soil food web. They range in size from the tiniest one-celled bacteria, algae, fungi, and protozoa, to the more complex nematodes and micro-arthropods, to the visible earthworms, insects, small vertebrates, and plants. As these organisms eat, grow, and move through the soil, they make it possible to have clean water, clean air, healthy plants, and moderated water flow.

There are many ways that the soil food web is an integral part of landscape processes. Soil organisms decompose organic compounds, including manure, plant residues, and pesticides, preventing them from entering water and becoming pollutants. They sequester nitrogen and other nutrients that might otherwise enter groundwater, and they fix nitrogen from the atmosphere, making it available to plants. Many organisms enhance soil aggregation and porosity, thus increasing infiltration and reducing surface runoff. Soil organisms prey on crop pests and are food for above-ground animals.

Research

Research interests span many aspects of soil ecology and microbiology, Fundamentally, researchers are interested in understanding the interplay among microorganisms, fauna, and plants, the biogeochemical processes they carry out, and the physical environment in which their activities take place, and applying this knowledge to address environmental problems.

Example research projects are to examine the biogeochemistry and microbial ecology of septic drain field soils used to treat domestic wastewater, the role of anecic earthworms in controlling the movement of water and nitrogen cycle in agricultural soils, and the assessment of soil quality in turf production.

Of particular interest as of 2006 is to understand the roles and functions of arbuscular mycorrhizal fungi in natural ecosystems. The effect of anthropic soil conditions on arbuscular mycorrhizal fungi, and the production of glomalin by arbuscular mycorrhizal fungi are both of particular interest due to their roles in sequestering atmospheric carbon dioxide.

Soil Biodiversity

Soil biodiversity refers both to the relationship of soil to biodiversity and to aspects of the soil that can be managed in relation to biodiversity. Soil biodiversity relates to some catchment management considerations.

Soil and Biodiversity

According to the Australian Department of the Environment and Water Resources, biodiversity is "the variety of life: the different plants, animals and micro-organisms, their genes and the ecosystems of which they are a part". Biodiversity and soil are strongly linked, because soil is the medium for a large variety of organisms, and interacts closely with the wider biosphere. Conversely, biological activity is a primary factor in the physical and chemical formation of soils.

Soil provides a vital habitat, primarily for microbes (including bacteria and fungi), but also for microfauna (such as protozoa and nematodes), mesofauna (such as microarthropods and enchytraeids), and macrofauna (such as earthworms, termites, and millipedes). The primary role of soil biota is to recycle organic matter that is derived from the "above-ground plant-based food web".

Soil is in close cooperation with the wider biosphere. The maintenance of fertile soil is "one of the most vital ecological services the living world performs", and the "mineral and organic contents of soil must be replenished constantly as plants consume soil elements and pass them up the food chain".

The correlation of soil and biodiversity can be observed spatially. For example, both natural and agricultural vegetation boundaries correspond closely to soil boundaries, even at continental and global scales.

Soil Management and Biodiversity

A "subtle synchrony" is how Baskin (1997) describes the relationship that exists between the soil and the diversity of life, above and below the ground. It is not surprising that soil management has a direct effect on biodiversity. This includes practices that influence soil volume, structure, biological, and chemical characteristics, and whether soil exhibits adverse effects such as reduced fertility, soil acidification, or salinisation. This section touches on selected soil factors that may be affected by soil management, and the according effect they can have on biodiversity.

Soil Process Effects

Soil Acidification

Soil acidity (or alkalinity) refers to the concentration of hydrogen ions (H^+) in the soil.

Measured on the pH scale, soil acidity is an invisible condition that directly affects soil fertility and toxicity by determining which elements in the soil are available for absorption by plants. Increases in soil acidity are caused by removal of agricultural product from the paddock, leaching of nitrogen as nitrate below the root zone, inappropriate use of nitrogenous fertilizers, and buildup of organic matter. Many of the soils in the Australian state of Victoria are naturally acidic; however, about 30,000 square kilometres or 23% of Victoria's agricultural soils suffer reduced productivity due to increased acidity. Soil acidity has been seen to damage the roots of the plants. Plants in higher acidity have smaller, less durable roots. Some evidence has shown that the acidity damages the tips of the roots restricting further growth. The height of the plants has also seen a marked restriction when grown in acidic soils, as seen in American and Russian wheat populations. The number of seeds that are even able to germinate in acidic soil is much lower than the amount of seeds that can sprout in a more neutral pH soil. These limitations to the growth of plants can have a very negative affect on plant health, leading to a decrease in overall plant population.

These effects occur regardless of the biome. A study in the Netherlands examined the correlation between soil pH and soil biodiversity in soils with pH below 5. A strong correlation was discovered, wherein the lower the pH the lower the biodiversity. The results were the same in grasslands as well as heathlands. Particularly concerning is the evidence showing that this acidification is directly linked to the decline in endangered species of plants, a trend recognized since 1950.

Soil acidification reduces soil biodiversity. It reduces the numbers of most macrofauna, including, for example, earthworm numbers (important in maintaining structural quality of the topsoil for plant growth). Also affected is rhizobium survival and persistence. Decomposition and nitrogen fixation may be reduced, which affects the survival of native vegetation. Biodiversity may further decline as certain weeds proliferate under declining native vegetation.

In strongly acidic soils, the associated toxicity may lead to decreased plant cover, leaving the soil susceptible to erosion by water and wind. Extremely low pH soils may suffer from structural decline as a result of reduced microrganisms and organic matter; this brings a susceptibility to erosion under high rainfall events, drought, and agricultural disturbance.

Some plants within the same species have shown resistance to the soil acidity their population grows in. Selectively breeding the stronger plants will be a way for humans to guard against the increasing soil acidity.

Further success combatting soil acidity has been seen in soybean and corn populations suffering from aluminum toxicity. Soil nutrients were restored and acidity decreased when lime was added to the soil. The plant health was increased and root biomass increased in response to the treatment. This is a possible solution for other acidic soil plant populations

Soil Structure Decline

Soil structure can be defined as the arrangement of particles and associated pores in soils across the size range from nanometres to centimeters. Biological influences can be demonstrated in the formation and stabilization of the soil aggregates, but it is necessary to distinguish clearly between those forces or agencies which create aggregations of particles and those which stabilize or degrade such aggregations. What qualifies as good soil contains the following attributes: optimal soil strength and aggregate stability, which offer resistance to structural degradation (capping/crusting, slaking and erosion, for example); optimal bulk density, which aids root development and contributes to other soil physical parameters such as water and air movement within the soil; optimal water holding capacity and rate of water infiltration.

Well-developed, healthy soils are complex systems in which physical soil structure is as important as chemical content. Soil pores—which are maximized in a well-structured soil—allow oxygen and moisture to infiltrate to depths and plant roots to penetrate to obtain moisture and nutrients.

Biological activity helps in the maintenance of relatively open soil structure, as well as facilitating decomposition and the transportation and transformation of soil nutrients. Changing soil structure has been shown to lead to reduced accessibility by plants to necessary substances. It is now uncontested that microbial exudates have a dominant role in the aggregation of soil particles and the protection of carbon from further degradation. It has been suggested that microorganisms within the soil "engineer" a superior habitat and provide a more sound soil structure, leading to more productive soil systems.

Traditional agricultural practices have generally caused declining soil structure. For example, cultivation causes the mechanical mixing of the soil, compacting and sheering of aggregates and filling of pore spaces - organic matter is also exposed to a greater rate of decay and oxidation. Soil structure is essential to soil health and fertility; soil structure decline has a direct effect on soil and surface food chain and biodiversity as a consequence. Continued crop cultivation eventually results in significant changes within the soil, such as its nutrient status, pH balance, organic matter content, and physical characteristics. While some of these changes can be beneficial to food and crop production, they can also be harmful towards other necessary systems. For example, studies have shown that tilling has had negative consequences towards soil organic matter (SOM), the organic component of soil composed of plant and animal decomposition and substances synthesized by soil organisms. SOM plays an integral role in the preservation of soil structure, but the constant tilling of crops has caused the SOM to shift and redistribute, causing soil structure to deteriorate and altering soil organism populations (such as with earthworms). Yet in many parts of the world, maximizing food production at all costs due to rampant poverty and the lack of food security tends to leave the long term ecological consequences overlooked, despite research and acknowledgment by the academic community.

Soil Sodicity

Soil sodicity refers the soil's content of sodium compared to its content of other cations, such as calcium. In high levels, sodium ions break apart clay platelets and cause swelling and dispersion in soil. This results in reduced soil sustainability. If the concentration occurs repeatedly, the soil will become cement-like, with little or no structure.

Extended exposure to high sodium levels results in a decrease in the amount of water retained and able to flow through soil, as well as a decrease in decomposition rates (this will leave the soil infertile and prohibit any future growth). This issue is prominent in Australia, where 1/3 of the land is affected by high levels of salt. It is a natural occurrence, but farming practices such as overgrazing and cultivation have contributed to the rise of it. The options for managing sodic soils are very limited; one must either change the plants or change the soil. The latter is the more difficult process. If changing the soil, one will need to add calcium to the soil to absorb the excess sodium blocking water flow.

Soil Salinisation

Soil salinity is the concentration of salt within the soil profile or on the soil surface. Excessive salt directly affects the composition of plants and animals due to varying salt tolerance – along with various physical and chemical changes to the soil, including structural decline and, in the extreme, denudation, exposure to soil erosion, and export of salts to waterways. At low soil salinity, there is a lot of microbial activity, that results in an increase in soil respiration, which increases the carbon dioxide levels in the soil, producing a healthier environment for plants. As the salinity of the soil rises, there is more stress on microbes because there is less available water available to them, leading to less respiration. Soil salinity has localised and regional effects on biodiversity, ranging, for example, from changes in plant composition and survival at a local discharge site through to regional changes in water quality and aquatic life.

While very saline soil is not preferred for growing crops, it is important to note that many crops can grow in more saline soils then others. This is important in countries where resources such as fresh water are scarce and needed for drinking, and saline water is able to be used for agriculture. Soil salinity can vary between extremes in a relatively small area; this allows plants to seek areas with less salinity. It is hard to determine which plants are able to grow in soil with high salinity, because the soil salinity is not uniform, even in small areas. However, it is found that plants will absorb nutrients from areas with lower salinity.

Soil Erosion

Soil erosion is the removal of the soil's upper layers by the action of water, wind, or ice. Soil erosion occurs naturally, but human activities can greatly increase its severity Soil that is healthy is fertile and productive. But soil erosion leads to a loss of topsoil, organic matter, and nutrients; it breaks down soil structure and decreases water storage capacity, in turn reducing fertility and the availability of water to plant roots. Soil erosion is therefore a major threat to soil biodiversity.

The effects of soil erosion can be lessened by means of various soil conservation techniques. These include changes in agricultural practice (such as moving to less erosion-prone crops), the planting of leguminous nitrogen-fixing trees, or trees that are known to replenish organic matter. Also, jute mats and jute geotextile nets can be used to divert and store runoff and control soil movement.

Misconstrued soil conservation efforts can result in an imbalance of soil chemical compounds. For example, attempts at afforestation in the northern Loess Plateau, China, have led to nutrient deprivation of organic materials such as carbon, nitrogen, and phosphorus. However, the right kind of species of plants is proven to improve an ecosystem if done properly.

Catchment Scale Impacts

Biological systems—both natural and artificial—depend heavily on healthy soils; it is the maintenance of soil health and fertility in all of its dimensions that sustains life. The interconnection spans vast spatial and temporal scales; the major degradation issues of salinity and soil erosion, for instance, can have anywhere from local to regional effects – it may take decades for the consequences of management actions affecting soil to be realised in terms of biodiversity impact.

Maintaining soil health is a regional or catchment-scale issue. Because soils are a dispersed asset, the only effective way is to ensure soil health generally is to encourage a broad, consistent, and economically appealing approach. Examples of such approaches as applied to an agricultural setting include the application of lime (calcium carbonate) to reduce acidity so to increase soil health and production, and the transition from conventional farming practices that employ cultivation to limited or no-till systems, which has had a positive impact on improving soil structure.

Conclusion

The relationship of soils to biodiversity is intimate and complex; it spans vast spatial and temporal scales and is essential to life. Soil is an asset as is biodiversity – the two should not be considered separately when it comes to protecting one or the other.

Soil can be managed to optimise its fertility and health under natural and agricultural land uses, so as to benefit biodiversity. Due to the dispersed nature of the soil asset, a broad but consistent and economically appealing approach to its protection is needed.

References

- Tóth, G., Stolbovoy, V. and Montanarella, 2007. Soil Quality and Sustainability Evaluation - An integrated approach to support soil-related policies of the European Union, EUR 22721 EN. 40 pp. Office for Official Publications of the European Communities, Luxembourg. ISBN 978-92-79-05250-7.

11

Processes of Soil Loss and Soil Contamination

Soil erosion and its prevention is a major area of study under soil science. The following chapter focuses on the varied reasons for soil contamination along with describing soil conservation processes as well. It also glances at land degradation with respect to soil. It will help the readers develop a deeper understanding of the phenomena of soil loss and contamination.

Erosion

An actively eroding rill on an intensively-farmed field in eastern Germany

In earth science, erosion is the action of surface processes (such as water flow or wind) that remove soil, rock, or dissolved material from one location on the Earth's crust, then transport it away to another location. The particulate breakdown of rock or soil into clastic sediment is referred to as *physical* or *mechanical* erosion; this contrasts with *chemical* erosion, where soil or rock material is removed from an area by its dissolving into a solvent (typically water), followed by the flow away of that solution.

Eroded sediment or solutes may be transported just a few millimetres, or for thousands of kilometres.

Natural rates of erosion are controlled by the action of geomorphic drivers, such as rainfall; bedrock wear in rivers; coastal erosion by the sea and waves; glacial plucking, abrasion, and scour; areal flooding; wind abrasion; groundwater processes; and mass movement processes in steep landscapes like landslides and debris flows. The rates at which such processes act control how fast a surface is eroded. Typically, physical erosion proceeds fastest on steeply sloping surfaces, and rates may also be sensitive to some climatically-controlled properties including amounts of water supplied (e.g., by rain), storminess, wind speed, wave fetch, or atmospheric temperature (especially for some ice-related processes). Feedbacks are also possible between rates of erosion and the amount of eroded material that is already carried by, for example, a river or glacier. Processes of erosion that produce sediment or solutes from a place contrast with those of deposition, which control the arrival and emplacement of material at a new location.

While erosion is a natural process, human activities have increased by 10-40 times the rate at which erosion is occurring globally. Excessive (or accelerated) erosion causes both "on-site" and "off-site" problems. On-site impacts include decreases in agricultural productivity and (on natural landscapes) ecological collapse, both because of loss of the nutrient-rich upper soil layers. In some cases, the eventual end result is desertification. Off-site effects include sedimentation of waterways and eutrophication of water bodies, as well as sediment-related damage to roads and houses. Water and wind erosion are the two primary causes of land degradation; combined, they are responsible for about 84% of the global extent of degraded land, making excessive erosion one of the most significant environmental problems world-wide.

Intensive agriculture, deforestation, roads, anthropogenic climate change and urban sprawl are amongst the most significant human activities in regard to their effect on stimulating erosion. However, there are many prevention and remediation practices that can curtail or limit erosion of vulnerable soils.

A natural arch produced by the wind erosion of differentially weathered rock in Jebel Kharaz, Jordan.

A wave-like sea cliff produced by coastal erosion, in Jinshitan Coastal National Geopark, Dalian, Liaoning Province, China.

Physical Processes

Rainfall and Surface Runoff

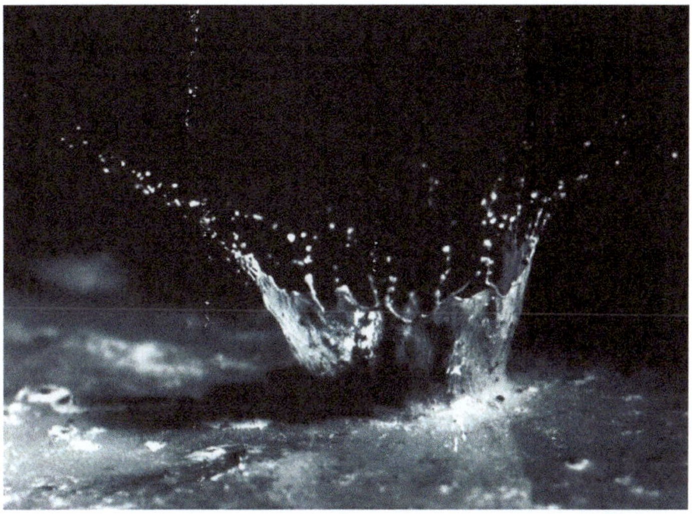

Soil and water being splashed by the impact of a single raindrop.

Rainfall, and the surface runoff which may result from rainfall, produces four main types of soil erosion: *splash erosion*, *sheet erosion*, *rill erosion*, and *gully erosion*. Splash erosion is generally seen as the first and least severe stage in the soil erosion process, which is followed by sheet erosion, then rill erosion and finally gully erosion (the most severe of the four).

In *splash erosion*, the impact of a falling raindrop creates a small crater in the soil, ejecting soil particles. The distance these soil particles travel can be as much as 0.6 m (two feet) vertically and 1.5 m (five feet) horizontally on level ground.

If the soil is saturated, or if the rainfall rate is greater than the rate at which water can infiltrate into the soil, surface runoff occurs. If the runoff has sufficient flow energy, it will transport loosened soil particles (sediment) down the slope. *Sheet erosion* is the transport of loosened soil particles by overland flow.

A spoil tip covered in rills and gullies due to erosion processes caused by rainfall: Rummu, Estonia

Rill erosion refers to the development of small, ephemeral concentrated flow paths which function as both sediment source and sediment delivery systems for erosion on hillslopes. Generally, where water erosion rates on disturbed upland areas are greatest, rills are active. Flow depths in rills are typically of the order of a few centimetres (about an inch) or less and along-channel slopes may be quite steep. This means that rills exhibit hydraulic physics very different from water flowing through the deeper, wider channels of streams and rivers.

Gully erosion occurs when runoff water accumulates and rapidly flows in narrow channels during or immediately after heavy rains or melting snow, removing soil to a considerable depth.

Rivers and Streams

Dobbingstone Burn, Scotland, showing two different types of erosion affecting the same place. Valley erosion is occurring due to the flow of the stream, and the boulders and stones (and much of the soil) that are lying on the stream's banks are glacial till that was left behind as ice age glaciers flowed over the terrain.

Valley or *stream erosion* occurs with continued water flow along a linear feature. The erosion is both downward, deepening the valley, and headward, extending the valley into the hillside, creating head cuts and steep banks. In the earliest stage of stream erosion, the erosive activity is dominantly vertical, the valleys have a typical V cross-section and the stream gradient is relatively steep. When some base level is reached, the erosive activity switches to lateral erosion, which widens the valley floor and creates a narrow floodplain. The stream gradient becomes nearly flat, and lateral deposition of sediments becomes important as the stream meanders across the valley floor. In all stages of stream erosion, by far the most erosion occurs during times of flood, when more and faster-moving water is available to carry a larger sediment load. In such processes, it is not the water alone that erodes: suspended abrasive particles, pebbles and boulders can also act erosively as they traverse a surface, in a process known as *traction*.

Bank erosion is the wearing away of the banks of a stream or river. This is distinguished from changes on the bed of the watercourse, which is referred to as *scour*. Erosion and changes in the form of river banks may be measured by inserting metal rods into the bank and marking the position of the bank surface along the rods at different times.

Thermal erosion is the result of melting and weakening permafrost due to moving water. It can occur both along rivers and at the coast. Rapid river channel migration observed in the Lena River of Siberia is due to thermal erosion, as these portions of the banks are composed of permafrost-cemented non-cohesive materials. Much of this erosion occurs as the weakened banks fail in large slumps. Thermal erosion also affects the Arctic coast, where wave action and near-shore temperatures combine to undercut permafrost bluffs along the shoreline and cause them to fail. Annual erosion rates along a 100-kilometre (62-mile) segment of the Beaufort Sea shoreline averaged 5.6 metres (18 feet) per year from 1955 to 2002.

Coastal Erosion

Wave cut platform caused by erosion of cliffs by the sea, at Southerndown in South Wales.

Erosion of the boulder clay (of Pleistocene age) along cliffs of Filey Bay, Yorkshire, England.

Shoreline erosion, which occurs on both exposed and sheltered coasts, primarily occurs through the action of currents and waves but sea level (tidal) change can also play a role.

Hydraulic action takes place when air in a joint is suddenly compressed by a wave closing the entrance of the joint. This then cracks it. *Wave pounding* is when the sheer energy of the wave hitting the cliff or rock breaks pieces off. *Abrasion* or *corrasion* is caused by waves launching seaload at the cliff. It is the most effective and rapid form of shoreline erosion. *Corrosion* is the dissolving of rock by carbonic acid in sea water. Limestone cliffs are particularly vulnerable to this kind of erosion. *Attrition* is where particles/seaload carried by the waves are worn down as they hit each other and the cliffs. This then makes the material easier to wash away. The material ends up as shingle and sand. Another significant source of erosion, particularly on carbonate coastlines, is the boring, scraping and grinding of organisms, a process termed *bioerosion*.

Sediment is transported along the coast in the direction of the prevailing current (longshore drift). When the upcurrent amount of sediment is less than the amount being carried away, erosion occurs. When the upcurrent amount of sediment is greater, sand or gravel banks will tend to form as a result of deposition. These banks may slowly migrate along the coast in the direction of the longshore drift, alternately protecting and exposing parts of the coastline. Where there is a bend in the coastline, quite often a buildup of eroded material occurs forming a long narrow bank (a spit). Armoured beaches and submerged offshore sandbanks may also protect parts of a coastline from erosion. Over the years, as the shoals gradually shift, the erosion may be redirected to attack different parts of the shore.

Chemical Erosion

Chemical erosion is the loss of matter in a landscape in the form of solutes. Chemical erosion is usually calculated from the solutes found in streams. Anders Rapp pioneered the study of chemical erosion in his work about Kärkevagge published in 1960.

Glaciers

Glacial moraines above Lake Louise, in Alberta, Canada.

Glaciers erode predominantly by three different processes: abrasion/scouring, plucking, and ice thrusting. In an abrasion process, debris in the basal ice scrapes along the bed, polishing and gouging the underlying rocks, similar to sandpaper on wood. Glaciers can also cause pieces of bedrock to crack off in the process of plucking. In ice thrusting, the glacier freezes to its bed, then as it surges forward, it moves large sheets of frozen sediment at the base along with the glacier. This method produced some of the many thousands of lake basins that dot the edge of the Canadian Shield. The erosion caused by glaciers worldwide erodes mountains so effectively that the term *glacial buzz-saw* has become widely used, which describes the limiting effect of glaciers on the height of mountain ranges. As mountains grow higher, they generally allow for more glacial activity (especially in the accumulation zone above the glacial equilibrium line altitude), which causes increased rates of erosion of the mountain, decreasing mass faster than isostatic rebound can add to the mountain. This provides a good example of a negative feedback loop. Ongoing research is showing that while glaciers tend to decrease mountain size, in some areas, glaciers can actually reduce the rate of erosion, acting as a *glacial armour*.

These processes, combined with erosion and transport by the water network beneath the glacier, leave moraines, drumlins, ground moraine (till), kames, kame deltas, moulins, and glacial erratics in their wake, typically at the terminus or during glacier retreat.

Floods

At extremely high flows, kolks, or vortices are formed by large volumes of rapidly rushing water. Kolks cause extreme local erosion, plucking bedrock and creating pothole-type geographical features called Rock-cut basins. Examples can be seen in the flood regions result from glacial Lake Missoula, which created the channeled scablands in the Columbia Basin region of eastern Washington.

Wind Erosion

Árbol de Piedra, a rock formation in the Altiplano, Bolivia sculpted by wind erosion.

Wind erosion is a major geomorphological force, especially in arid and semi-arid regions. It is also a major source of land degradation, evaporation, desertification, harmful airborne dust, and crop damage—especially after being increased far above natural rates by human activities such as deforestation, urbanization, and agriculture.

Wind erosion is of two primary varieties: *deflation*, where the wind picks up and carries away loose particles; and *abrasion*, where surfaces are worn down as they are struck by airborne particles carried by wind. Deflation is divided into three categories: (1) *surface creep*, where larger, heavier particles slide or roll along the ground; (2) *saltation*, where particles are lifted a short height into the air, and bounce and saltate across the surface of the soil; and (3) *suspension*, where very small and light particles are lifted into the air by the wind, and are often carried for long distances. Saltation is responsible for the majority (50-70%) of wind erosion, followed by suspension (30-40%), and then surface creep (5-25%).

Wind erosion is much more severe in arid areas and during times of drought. For example, in the Great Plains, it is estimated that soil loss due to wind erosion can be as much as 6100 times greater in drought years than in wet years.

Mass Movement

Wadi in Makhtesh Ramon, Israel, showing gravity collapse erosion on its banks.

Mass movement is the downward and outward movement of rock and sediments on a sloped surface, mainly due to the force of gravity.

Mass movement is an important part of the erosional process, and is often the first stage in the breakdown and transport of weathered materials in mountainous areas. It moves material from higher elevations to lower elevations where other eroding agents such as streams and glaciers can then pick up the material and move it to even lower elevations. Mass-movement processes are always occurring continuously on all slopes; some mass-movement processes act very slowly; others occur very suddenly, often with disastrous results. Any perceptible down-slope movement of rock or sediment is often referred to in general terms as a landslide. However, landslides can be classified in a much more detailed way that reflects the mechanisms responsible for the movement and the velocity at which the movement occurs. One of the visible topographical manifestations of a very slow form of such activity is a scree slope.

Slumping happens on steep hillsides, occurring along distinct fracture zones, often within materials like clay that, once released, may move quite rapidly downhill. They will often show a spoon-shaped isostatic depression, in which the material has begun to slide downhill. In some cases, the slump is caused by water beneath the slope weakening it. In many cases it is simply the result of poor engineering along highways where it is a regular occurrence.

Surface creep is the slow movement of soil and rock debris by gravity which is usually not perceptible except through extended observation. However, the term can also describe the rolling of dislodged soil particles 0.5 to 1.0 mm (0.02 to 0.04 in) in diameter by wind along the soil surface.

Factors Affecting Erosion Rates

Climate

The amount and intensity of precipitation is the main climatic factor governing soil erosion by water. The relationship is particularly strong if heavy rainfall occurs at times when, or in locations where, the soil's surface is not well protected by vegetation. This might be during periods when agricultural activities leave the soil bare, or in semi-arid regions where vegetation is naturally sparse. Wind erosion requires strong winds, particularly during times of drought when vegetation is sparse and soil is dry (and so is more erodible). Other climatic factors such as average temperature and temperature range may also affect erosion, via their effects on vegetation and soil properties. In general, given similar vegetation and ecosystems, areas with more precipitation (especially high-intensity rainfall), more wind, or more storms are expected to have more erosion.

In some areas of the world (e.g. the mid-western USA), rainfall intensity is the primary determinant of erosivity, with higher intensity rainfall generally resulting in more soil erosion by water. The size and velocity of rain drops is also an important factor. Larger and higher-velocity rain drops have greater kinetic energy, and thus their impact will displace soil particles by larger distances than smaller, slower-moving rain drops.[1]

In other regions of the world (e.g. western Europe), runoff and erosion result from relatively low intensities of stratiform rainfall falling onto previously saturated soil. In such situations, rainfall amount rather than intensity is the main factor determining the severity of soil erosion by water.

Vegetative Cover

Vegetation acts as an interface between the atmosphere and the soil. It increases the permeability of the soil to rainwater, thus decreasing runoff. It shelters the soil from winds, which results in decreased wind erosion, as well as advantageous changes in microclimate. The roots of the plants bind the soil together, and interweave with other roots, forming a more solid mass that is less susceptible to both water and wind erosion. The removal of vegetation increases the rate of surface erosion.

Topography

The topography of the land determines the velocity at which surface runoff will flow, which in turn determines the erosivity of the runoff. Longer, steeper slopes (especially those without adequate vegetative cover) are more susceptible to very high rates of erosion during heavy rains than shorter, less steep slopes. Steeper terrain is also more prone to mudslides, landslides, and other forms of gravitational erosion processes.

Tectonics

Tectonic processes control rates and distributions of erosion at the Earth's surface. If tectonic action causes part of the Earth's surface (e.g., a mountain range) to be raised or lowered relative to surrounding areas, this must necessarily change the gradient of the land surface. Because erosion rates are almost always sensitive to local slope, this will change the rates of erosion in the uplifted area. Active tectonics also brings fresh, unweathered rock towards the surface, where it is exposed to the action of erosion.

However, erosion can also affect tectonic processes. The removal by erosion of large amounts of rock from a particular region, and its deposition elsewhere, can result in a lightening of the load on the lower crust and mantle. Because tectonic processes are driven by gradients in the stress field developed in the crust, this unloading can in turn cause tectonic or isostatic uplift in the region. In some cases, it has been hypothesised that these twin feedbacks can act to localise and enhance zones of very rapid exhumation of deep crustal rocks beneath places on the Earth's surface with extremely high erosion rates, for example, beneath the extremely steep terrain of Nanga Parbat in the western Himalayas. Such a place has been called a "tectonic aneurysm".

Erosion of Earth Systems

Mountain Ranges

Mountain ranges are known to take many million of years to erode to the degree they effectively cease to exist. Scholars Pitman and Golovchenko estimate that it takes probably more than 450 million years to erode a mountain mass similar to the Himalaya into an almost-flat peneplain if there are no major sea-level changes. Erosion of mountains massifs can create a pattern of equally high summits called summit accordance.

Examples of heavily eroded mountain ranges include the Timanides of Northern Russia. Erosion of this orogen has produced sediments that are now found in the East European Platform, including the Cambrian Sablya Formation near Lake Ladoga. Studies of these sediments points that its likely that the erosion of the orogen was beginning in the Cambrian and then became stronger in Ordovician.

Soils

If the rate of erosion is higher than the rate of soil formation the soils are being destroyed by erosion. Where soil is not destroyed by erosion, erosion can in some cases prevent the formation of soil features that form slowly. Inceptisols are common soils that form in areas of fast erosion.

While erosion of soils is a natural process, human activities have increased by 10-40 times the rate at which erosion is occurring globally. Excessive (or accelerated) erosion causes both "on-site" and "off-site" problems. On-site impacts include decreases in agricultural productivity and (on natural landscapes) ecological collapse, both because of loss of the nutrient-rich upper soil layers. In some cases, the eventual end result is desertification. Off-site effects include sedimentation of waterways and eutrophication of water bodies, as well as sediment-related damage to roads and houses. Water and wind erosion are the two primary causes of land degradation; combined, they are responsible for about 84% of the global extent of degraded land, making excessive erosion one of the most significant environmental problems world-wide.

Prevention of Soil Loss

Soil Conservation

Soil conservation is the prevention of soil loss from erosion or reduced fertility caused by over usage, acidification, salinization or other chemical soil contamination.

Slash-and-burn and other unsustainable methods of subsistence farming are practiced in some lesser developed areas. A sequel to the deforestation is typically large scale erosion, loss of soil nutrients and sometimes total desertification.

Techniques for improved soil conservation include crop rotation, cover crops, conservation tillage and planted windbreaks and affect both erosion and fertility. When plants, especially trees, die, they decay and become part of the soil. Code 330 defines standard methods recommended by the US Natural Resources Conservation Service.

Erosion barriers on disturbed slope, Marin County, California

Contour plowing, Pennsylvania 1938. The rows formed slow water run-off during rainstorms to prevent soil erosion and allows the water time to infiltrate into the soil.

Farmers have practiced soil conservation for millennia. Conservation practices fall in multiple categories:

Contour Plowing

Contour plowing orients crop furrows following the contour lines of the farmed area. Furrows move left and right to maintain a constant altitude, which reduces runoff. Contour plowing was practiced by the ancient Phoenicians, and is effective for slopes between two and ten percent. Contour plowing can increase crop yields from 10 to 50 percent, partially as a result from greater soil retention.

Terracing or Terrace Farming

Terracing is the practice of creating nearly level areas in a hillside area. The terraces form a series of steps, each at a higher level than the previous. Terraces are protected from erosion by other soil barriers.

Terraced farming is more common on small farms and in underdeveloped countries, since mechanized equipment is difficult to deploy in this setting.

Keyline Design

Keyline design is an enhancement of contour farming, where the total watershed properties are taken into account in forming the contour lines.

Perimeter Runoff Control

Tree, shrubs and ground-cover are effective perimeter treatment for soil erosion prevention, by impeding surface flows. A special form of this perimeter or inter-row treatment is the use of a "grass way" that both channels and dissipates runoff through surface friction, impeding surface runoff and encouraging infiltration of the slowed surface water.

Windbreaks

Windbreaks are sufficiently dense rows of trees at the windward exposure of an agricultural field subject to wind erosion. Evergreen species provide year-round protection; however, as long as foliage is present in the seasons of bare soil surfaces, the effect of deciduous trees may be adequate.

Cover Crops/Crop Rotation

Cover crops such as legumes, white turnip, radishes and other species are rotated with cash crops to blanket the soil year-round and act as green manure that replenishes nitrogen and other critical nutrients. Cover crops also help suppress weeds.

Soil-Conservation Farming

Soil-conservation farming involves no-till farming, "green manures" and other soil-enhancing practices. Such farming methods attempt to mimic the biology of virgin land. They can revive damaged soil, minimize erosion, encourage plant growth, eliminate the use of nitrogen fertilizer or fungicide, produce above-average yields and protect crops during droughts or flooding. The result is less labor and lower costs that increase farmers' profits. No-till farming and cover crops act as sinks for nitrogen and other nutrients. This increases the amount of soil organic matter.

Repeated plowing/tilling degrades soil, killing its beneficial fungi and earthworms. Once damaged, soil may take multiple seasons to fully recover, even in optimal circumstances.

Critics argue that no-till and related methods are impractical and too expensive for many growers, partly because it requires new equipment. They cite advantages for conventional tilling depending on the geography, crops and soil conditions. Some farmers claimed that no-till complicates weed control, delays planting and that post-harvest residues, especially for corn, are hard to manage.

Salinity Management

Salt deposits on the former bed of the Aral Sea

Salinity in soil is caused by irrigating with salty water. Water then evaporates from the soil leaving the salt behind. Salt breaks down the soil structure, causing infertility and reduced growth.

The ions responsible for salination are: sodium (Na^{+}), potassium (K^{+}), calcium (Ca^{2+}), magnesium (Mg^{2+}) and chlorine (Cl^{-}). Salinity is estimated to affect about one third of the earth's arable land. Soil salinity adversely affects crop metabolism and erosion usually follows.

Salinity occurs on drylands from overirrigation and in areas with shallow saline water tables. Over-irrigation deposits salts in upper soil layers as a byproduct of soil infiltration; irrigation merely increases the rate of salt deposition. The best-known case of shallow saline water table capillary action occurred in Egypt after the 1970 construction of the Aswan Dam. The change in the groundwater level led to high salt concentrations in the water table. The continuous high level of the water table led to soil salination.

Use of humic acids may prevent excess salination, especially given excessive irrigation. Humic acids can fix both anions and cations and eliminate them from root zones.

Planting species that can tolerate saline conditions to produce surface cover can preserve soil salinity can be reduced. Salt-tolerant plants include saltbush, a plant found in much of North America and in the Mediterranean regions of Europe.

Soil Organisms

Yellow fungus, a mushroom that assists in organic decay.

When worms excrete egesta in the form of casts, a balanced selection of minerals and plant nutrients is made into a form accessible for root uptake. Earthworm casts are five times richer in available nitrogen, seven times richer in available phosphates and eleven times richer in available potash than the surrounding upper 150 millimetres (5.9 in) of soil. The weight of casts produced may be greater than 4.5 kg per worm per year. By burrowing, the earthworm improves soil porosity, creating channels that enhance the processes of aeration and drainage.

Other important soil organisms include nematodes, mycorrizha and bacteria.

Degraded soil requires synthetic fertilizer to produce high yields. Lacking structure increases erosion and carries nitrogen and other pollutants into rivers and streams.

Each one percent increase in soil organic matter helps soil hold 20,000 gallons more water per acre.

Mineralization

To allow plants full realization of their phytonutrient potential, active mineralization of the soil is sometimes undertaken. This can involve adding crushed rock or chemical soil supplements. In either case the purpose is to combat mineral depletion. A broad range of minerals can be used, including common substances such as phosphorus and more exotic substances such as zinc and selenium. Extensive research examines the phase transitions of minerals in soil with aqueous contact.

Flooding can bring significant sediment to an alluvial plain. While this effect may not be desirable if floods endanger life or if the sediment originates from productive land, this process of addition to a floodplain is a natural process that can rejuvenate soil chemistry through mineralization and macronutrient addition.

US

For some crops, no-tillage acreage nearly doubled from 2000 to 2015. For soybeans, for example, it rose to 30 million acres in 2012 from 16.5 million acres in 1996. Use of cover crops was rising in acreage about 30 percent a year, according to surveys, though the total remains small.

Soil Liquefaction

Some effects of liquefaction after the 1964 Niigata earthquake

Liquefaction allowed this sewer to float upward – 2004 Chūetsu earthquake

The effect of liquefaction in Christchurch, New Zealand, during the Mw 6.3 February 2011 Christchurch earthquake

Soil liquefaction describes a phenomenon whereby a saturated or partially saturated soil substantially loses strength and stiffness in response to an applied stress, usually earthquake shaking or other sudden change in stress condition, causing it to behave like a liquid.

In soil mechanics the term "liquefied" was first used by Allen Hazen in reference to the 1918 failure of the Calaveras Dam in California. He described the mechanism of flow liquefaction of the embankment dam as follows:

If the pressure of the water in the pores is great enough to carry all the load, it will have the effect of holding the particles apart and of producing a condition that is practically equivalent to that of quicksand... the initial movement of some part of the material might result in accumulating pressure, first on one point, and then on another, successively, as the early points of concentration were liquefied.

The phenomenon is most often observed in saturated, loose (low density or uncompacted), sandy soils. This is because a loose sand has a tendency to compress when a load is applied; dense sands by contrast tend to expand in volume or 'dilate'. If the soil is saturated by water, a condition that often exists when the soil is below the ground water table or sea level, then water fills the gaps between soil grains ('pore spaces'). In response to the soil compressing, this water increases in pressure and attempts to flow out from the soil to zones of low pressure (usually upward towards the ground surface). However, if the loading is rapidly applied and large enough, or is repeated many times (e.g. earthquake shaking, storm wave loading) such that it does not flow out in time before the next cycle of load is applied, the water pressures may build to an extent where they exceed the contact stresses between the grains of soil that keep them in contact with each other. These contacts between grains are the means by which the weight from buildings and overlying soil layers are transferred from the ground surface to layers of soil or rock at greater depths. This loss of soil structure causes it to lose all of its strength (the ability to transfer shear stress) and it may be observed to flow like a liquid (hence 'liquefaction').

Although the effects of liquefaction have been long understood, it was more thoroughly brought to the attention of engineers after the 1964 Niigata earthquake and 1964 Alaska earthquake. It was also a major factor in the destruction in San Francisco's Marina District during the 1989 Loma Prieta earthquake, and in Port of Kobe during the 1995 Great Hanshin earthquake. More recently liquefaction was largely responsible for extensive damage to residential properties in the eastern suburbs and satellite townships of Christchurch, New Zealand during the 2010 Canterbury earthquake and more extensively again following the Christchurch earthquakes that followed in early and mid-2011.

The building codes in many developed countries require engineers to consider the effects of soil liquefaction in the design of new buildings and infrastructure such as bridges, embankment dams and retaining structures.

Technical Definitions

A state of 'soil liquefaction' occurs when the effective stress of soil is reduced to essentially zero, which corresponds to a complete loss of shear strength. This may be initiated by either monotonic loading (e.g. single sudden occurrence of a change in stress – examples include an increase in load on an embankment or sudden loss of toe support) or cyclic loading (e.g. repeated change in stress condition – examples include wave loading or earthquake shaking). In both cases a soil in a saturated loose state, and one which may generate significant pore water pressure on a change in load are the most likely to liquefy. This is because a loose soil has the tendency to compress when sheared, generating large excess porewater pressure as load is transferred from the soil skeleton to adjacent pore water during undrained loading. As pore water pressure rises a progressive loss of strength of the soil occurs as effective stress is reduced. It is more likely to occur in sandy or non-plastic silty soils, but may in rare cases occur in gravels and clays.

A 'flow failure' may initiate if the strength of the soil is reduced below the stresses required to maintain equilibrium of a slope or footing of a building for instance. This can occur due to monotonic loading or cyclic loading, and can be sudden and catastrophic. A historical example is the Aberfan disaster. Casagrande referred to this type of phenomena as 'flow liquefaction' although a state of zero effective stress is not required for this to occur.

The term 'cyclic liquefaction' refers to the occurrence of a state of soil when large shear strains have accumulated in response to cyclic loading. A typical reference strain for the approximate occurrence of zero effective stress is 5% double amplitude shear strain. This is a soil test based definition, usually performed via cyclic triaxial, cyclic direct simple shear, or cyclic torsional shear type apparatus. These tests are performed to determine a soil's resistance to liquefaction by observing the number of cycles of loading at a particular shear stress amplitude before it 'fails'. Failure here is defined by the aforementioned shear strain criteria.

The term 'cyclic mobility' refers to the mechanism of progressive reduction of effective stress due to cyclic loading. This may occur in all soil types including dense soils. However, on reaching a state of zero effective stress such soils immediate dilate and regain strength. Thus shear strains are significantly less than a true state of soil liquefaction whereby a loose soil exhibits flow type phenomena.

Occurrence

Liquefaction is more likely to occur in loose to moderately saturated granular soils with poor drainage, such as silty sands or sands and gravels capped or containing seams of impermeable sediments. During wave loading, usually cyclic undrained loading, e.g. seismic loading, loose sands tend to decrease in volume, which produces an increase in their pore water pressures and consequently a decrease in shear strength, i.e. reduction in effective stress.

Deposits most susceptible to liquefaction are young (Holocene-age, deposited within the last 10,000 years) sands and silts of similar grain size (well-sorted), in beds at least metres thick, and saturated with water. Such deposits are often found along stream beds, beaches, dunes, and areas where windblown silt (loess) and sand have accumulated. Some examples of soil liquefaction include quicksand, quick clay, turbidity currents, and earthquake induced liquefaction.

Depending on the initial void ratio, the soil material can respond to loading either *strain-softening* or *strain-hardening*. Strain-softened soils, e.g. loose sands, can be triggered to collapse, either monotonically or cyclically, if the static shear stress is greater than the ultimate or steady-state shear strength of the soil. In this case *flow liquefaction* occurs, where the soil deforms at a low constant residual shear stress. If the soil strain-hardens, e.g. moderately dense to dense sand, flow liquefaction will generally not occur. However, *cyclic softening* can occur due to cyclic undrained loading, e.g. earthquake loading. Deformation during cyclic loading will depend on the density of the soil, the magnitude and duration of the cyclic loading, and amount of shear stress reversal. If stress reversal occurs, the effective shear stress could reach zero, then *cyclic liquefaction* can take place. If stress reversal does not occur, zero effective stress is not possible to occur, then *cyclic mobility* takes place.

The resistance of the cohesionless soil to liquefaction will depend on the density of the soil, confining stresses, soil structure (fabric, age and cementation), the magnitude and duration of the cyclic loading, and the extent to which shear stress reversal occurs.

Earthquake Liquefaction

Sand boils that erupted during the 2011 Christchurch earthquake.

The pressures generated during large earthquakes with many cycles of shaking can cause the liquefied sand and excess water to force its way to the ground surface from several metres below the ground. This is often observed as "sand boils" also called "sand blows" or "sand volcanoes" (as they appear to form small volcanic craters) at the ground surface. The phenomenon may incorporate both flow of already liquefied sand from a layer below ground, and a quicksand effect whereby upward flow of water initiates liquefaction in overlying non-liquefied sandy deposits due to buoyancy.

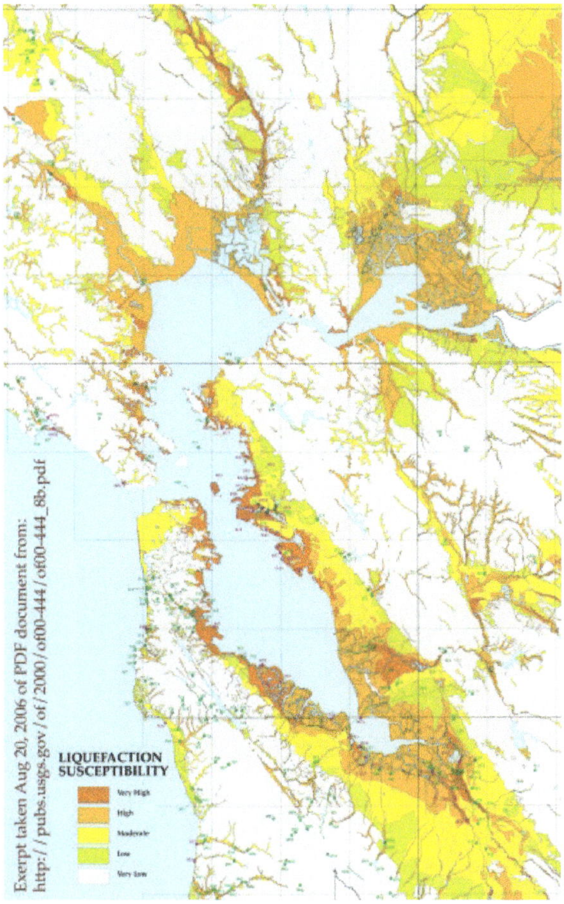

A liquefaction susceptibility map – excerpt of USGS map for the San Francisco Bay Area. Many areas of concern in this region are also densely urbanized.

The other common observation is land instability – cracking and movement of the ground down slope or towards unsupported margins of rivers, streams, or the coast. The failure of ground in this manner is called 'lateral spreading', and may occur on very shallow slopes of angles of only 1 or 2 degrees from the horizontal.

One positive aspect of soil liquefaction is the tendency for the effects of earthquake shaking to be significantly damped (reduced) for the remainder of the earthquake. This is because liquids do not support a shear stress and so once the soil liquefies due to shaking, subsequent earthquake shaking (transferred through ground by shear waves) is not transferred to buildings at the ground surface.

Studies of liquefaction features left by prehistoric earthquakes, called paleoliquefaction or paleoseismology, can reveal a great deal of information about earthquakes that occurred before records were kept or accurate measurements could be taken.

Soil liquefaction induced by earthquake shaking is also a major contributor to urban seismic risk.

Effects

The effects of lateral spreading (River Road in Christchurch following the 2011 Christchurch earthquake)

Damage in Brooklands from the 2010 Canterbury earthquake, where buoyancy caused by soil liquefaction pushed up an underground service including this manhole

The effects of soil liquefaction on the built environment can be extremely damaging. Buildings whose foundations bear directly on sand which liquefies will experience a sudden loss of support, which will result in drastic and irregular settlement of the building causing structural damage, including cracking of foundations and damage to the building structure itself, or may leave the structure unserviceable afterwards, even without structural damage. Where a thin crust of non-liquefied soil exists between building foundation and liquefied soil, a 'punching shear' type foundation failure may occur. The irregular settlement of ground may also break underground utility lines. The upward pressure applied by the movement of liquefied soil through the crust layer can crack weak foundation slabs and enter buildings through service ducts, and may allow water to damage the building contents and electrical services.

Bridges and large buildings constructed on pile foundations may lose support from the adjacent soil and buckle, or come to rest at a tilt after shaking.

Sloping ground and ground next to rivers and lakes may slide on a liquefied soil layer (termed 'lateral spreading'), opening large cracks or fissures in the ground, and can cause significant damage to buildings, bridges, roads and services such as water, natural gas, sewerage, power and telecommunications installed in the affected ground. Buried tanks and manholes may float in the liquefied soil due to buoyancy. Earth embankments such as flood levees and earth dams may lose stability or collapse if the material comprising the embankment or its foundation liquefies.

Mitigation Methods

Methods to mitigate the effects of soil liquefaction have been devised by earthquake engineers and include various soil compaction techniques such as vibro compaction (compaction of the soil by depth vibrators), dynamic compaction, and vibro stone columns. These methods result in the densification of soil and enable buildings to withstand soil liquefaction.

Existing buildings can be mitigated by injecting grout into the soil to stabilize the layer of soil that is subject to liquefaction.

Quicksand

Quicksand forms when water saturates an area of loose sand and the ordinary sand is agitated. When the water trapped in the batch of sand cannot escape, it creates liquefied soil that can no longer support weight. Quicksand can be formed by standing or (upwards) flowing underground water (as from an underground spring), or by earthquakes. In the case of flowing underground water, the force of the water flow opposes the force of gravity, causing the granules of sand to be more buoyant. In the case of earthquakes, the shaking force can increase the pressure of shallow groundwater, liquefying sand and silt deposits. In both cases, the liquefied surface loses strength, causing buildings or other objects on that surface to sink or fall over.

The saturated sediment may appear quite solid until a change in pressure or shock initiates the liquefaction, causing the sand to form a suspension with each grain surrounded by a thin film of water. This cushioning gives quicksand, and other liquefied sediments, a spongy, fluidlike texture. Objects in the liquefied sand sink to the level at which the weight of the object is equal to the weight of the displaced sand/water mix and the object *floats* due to its buoyancy.

Quick clay

Quick clay, also known as *Leda Clay* in Canada, is a water-saturated gel, which in its solid form resemble a unique form of highly sensitive clay. This clay has a tendency to change from a relatively stiff condition to a liquid mass when it is disturbed. This gradual change in appearance from solid to liquid is a process known as spontaneous liquefaction. The clay retains a solid structure despite the high water content (up to 80% by volume), because surface tension holds water-coated flakes of clay together in a delicate structure. When the structure is broken by a shock or sufficient shear, it turns to a fluid state.

Quick clay is only found in the northern countries such as Russia, Canada, Alaska in the U.S., Norway, Sweden, and Finland, which were glaciated during the Pleistocene epoch.

Quick clay has been the underlying cause of many deadly landslides. In Canada alone, it has been associated with more than 250 mapped landslides. Some of these are ancient, and may have been triggered by earthquakes.

Turbidity Currents

Submarine landslides are turbidity currents and consist of water saturated sediments flowing downslope. An example occurred during the 1929 Grand Banks earthquake that struck the continental slope off the coast of Newfoundland. Minutes later, transatlantic telephone cables began breaking sequentially, farther and farther downslope, away from the epicenter. Twelve cables were snapped in a total of 28 places. Exact times and locations were recorded for each break. Investigators suggested that a 60-mile-per-hour (100 km/h) submarine *landslide* or turbidity current of water saturated sediments swept 400 miles (600 km) down the continental slope from the earthquake's epicenter, snapping the cables as it passed.

Soil Contamination

Soil contamination or soil pollution as part of land degradation is caused by the presence of xenobiotic (human-made) chemicals or other alteration in the natural soil environment. It is typically caused by industrial activity, agricultural chemicals, or improper disposal of waste. The most common chemicals involved are petroleum hydrocarbons, polynuclear aromatic hydrocarbons (such as naphthalene and benzo(a) pyrene), solvents, pesticides, lead, and other heavy metals. Contamination is correlated with the degree of industrialization and intensity of chemical usage.

The concern over soil contamination stems primarily from health risks, from direct contact with the contaminated soil, vapors from the contaminants, and from secondary contamination of water supplies within and underlying the soil. Mapping of contaminated soil sites and the resulting cleanup are time consuming and expensive tasks, requiring extensive amounts of geology, hydrology, chemistry, computer modeling skills, and GIS in Environmental Contamination, as well as an appreciation of the history of industrial chemistry.

Excavation showing soil contamination at a disused gasworks

In North America and Western Europe the extent of contaminated land is best known, with many of countries in these areas having a legal framework to identify and deal with this environmental problem. Developing countries tend to be less tightly regulated despite some of them having undergone significant industrialization.

Causes

Soil pollution can be caused by the following (non-exhaustive list!):

- Oil drilling
- Mining and activities by other heavy industries
- Accidental spills as may happen during deforestation activities, etc.
- Corrosion of underground storage tanks (including piping used to transmit the contents)
- Acid rain (in turn caused by air pollution)
- Intensive farming
- Agrochemicals, such as pesticides, herbicides and fertilizers
- Industrial accidents

- Road debris
- Drainage of contaminated surface water into the soil
- Waste disposal
  - Oil and fuel dumping
  - Nuclear wastes
  - Direct discharge of industrial wastes to the soil
  - Landfill and illegal dumping
  - coal ash
  - Electronic waste
  - ammunitions and agents of war

The most common chemicals involved are petroleum hydrocarbons, solvents, pesticides, lead, and other heavy metals.

In a wider sense, genetically modified plants (GMP) can count as a risk factor for soils, because of their potential to affect the soil fauna. Any activity that leads to other forms of soil degradation (erosion, compaction, etc.) may indirectly worsen the contamination effects in that soil remediation becomes more tedious.

Historical deposition of coal ash used for residential, commercial, and industrial heating, as well as for industrial processes such as ore smelting, were a common source of contamination in areas that were industrialized before about 1960. Coal naturally concentrates lead and zinc during its formation, as well as other heavy metals to a lesser degree. When the coal is burned, most of these metals become concentrated in the ash (the principal exception being mercury). Coal ash and slag may contain sufficient lead to qualify as a "characteristic hazardous waste", defined in the USA as containing more than 5 mg/l of extractable lead using the TCLP procedure. In addition to lead, coal ash typically contains variable but significant concentrations of polynuclear aromatic hydrocarbons (PAHs; e.g., benzo(a)anthracene, benzo(b)fluoranthene, benzo(k)fluoranthene, benzo(a)pyrene, indeno(cd)pyrene, phenanthrene, anthracene, and others). These PAHs are known human carcinogens and the acceptable concentrations of them in soil are typically around 1 mg/kg. Coal ash and slag can be recognised by the presence of off-white grains in soil, gray heterogeneous soil, or (coal slag) bubbly, vesicular pebble-sized grains.

Treated sewage sludge, known in the industry as biosolids, has become controversial as a "fertilizer". As it is the byproduct of sewage treatment, it generally contains more contaminants such as organisms, pesticides, and heavy metals than other soil.

In the European Union, the Urban Waste Water Treatment Directive allows sewage sludge to be sprayed onto land. The volume is expected to double to 185,000 tons of dry solids in 2005. This has good agricultural properties due to the high nitrogen and phosphate content. In 1990/1991, 13% wet weight was sprayed onto 0.13% of the land; however, this is expected to rise 15 fold by 2005. Advocates say there is a need to control this so that pathogenic microorganisms do not get into water courses and to ensure that there is no accumulation of heavy metals in the top soil.

Pesticides and Herbicides

A pesticide is a substance or mixture of substances used to kill a pest. A pesticide may be a chemical substance, biological agent (such as a virus or bacteria), antimicrobial, disinfectant or device used against any pest. Pests include insects, plant pathogens, weeds, mollusks, birds, mammals, fish, nematodes (roundworms) and microbes that compete with humans for food, destroy property, spread or are a vector for disease or cause a nuisance. Although there are benefits to the use of pesticides, there are also drawbacks, such as potential toxicity to humans and other organisms.

Herbicides are used to kill weeds, especially on pavements and railways. They are similar to auxins and most are biodegradable by soil bacteria. However, one group derived from trinitrotoluene (2:4 D and 2:4:5 T) have the impurity dioxin, which is very toxic and causes fatality even in low concentrations. Another herbicide is Paraquat. It is highly toxic but it rapidly degrades in soil due to the action of bacteria and does not kill soil fauna.

Insecticides are used to rid farms of pests which damage crops. The insects damage not only standing crops but also stored ones and in the tropics it is reckoned that one third of the total production is lost during food storage. As with fungicides, the first insecticides used in the nineteenth century were inorganic e.g. Paris Green and other compounds of arsenic. Nicotine has also been used since the late eighteenth century.

There are now two main groups of synthetic insecticides -

1. Organochlorines include DDT, Aldrin, Dieldrin and BHC. They are cheap to produce, potent and persistent. DDT was used on a massive scale from the 1930s, with a peak of 72,000 tonnes used 1970. Then usage fell as the harmful environmental effects were realized. It was found worldwide in fish and birds and was even discovered in the snow in the Antarctic. It is only slightly soluble in water but is very soluble in the bloodstream. It affects the nervous and endocrine systems and causes the eggshells of birds to lack calcium causing them to be easily breakable. It is thought to be responsi-

ble for the decline of the numbers of birds of prey like ospreys and peregrine falcons in the 1950s - they are now recovering. As well as increased concentration via the food chain, it is known to enter via permeable membranes, so fish get it through their gills. As it has low water solubility, it tends to stay at the water surface, so organisms that live there are most affected. DDT found in fish that formed part of the human food chain caused concern, but the levels found in the liver, kidney and brain tissues was less than 1 ppm and in fat was 10 ppm, which was below the level likely to cause harm. However, DDT was banned in the UK and the United States to stop the further buildup of it in the food chain. U.S. manufacturers continued to sell DDT to developing countries, who could not afford the expensive replacement chemicals and who did not have such stringent regulations governing the use of pesticides..

2. Organophosphates, e.g. parathion, methyl parathion and about 40 other insecticides are available nationally. Parathion is highly toxic, methyl-parathion is less so and Malathion is generally considered safe as it has low toxicity and is rapidly broken down in the mammalian liver. This group works by preventing normal nerve transmission as cholinesterase is prevented from breaking down the transmitter substance acetylcholine, resulting in uncontrolled muscle movements.

Agents of War

The disposal of munitions, and a lack of care in manufacture of munitions caused by the urgency of production, can contaminate soil for extended periods. There is little published evidence on this type of contamination largely because of restrictions placed by Governments of many countries on the publication of material related to war effort. However, mustard gas stored during World War II has contaminated some sites for up to 50 years and the testing of Anthrax as a potential biological weapon contaminated the whole island of Gruinard

Health Effects

Contaminated or polluted soil directly affects human health through direct contact with soil or via inhalation of soil contaminants which have vaporized; potentially greater threats are posed by the infiltration of soil contamination into groundwater aquifers used for human consumption, sometimes in areas apparently far removed from any apparent source of above ground contamination.

Health consequences from exposure to soil contamination vary greatly depending on pollutant type, pathway of attack and vulnerability of the exposed population. Chronic exposure to chromium, lead and other metals, petroleum, solvents, and many pesticide and herbicide formulations can be carcinogenic, can cause congenital disorders, or can cause other chronic health conditions. Industrial or man-made concentrations of naturally occurring substances, such as nitrate and ammonia associated with livestock manure from agricultural operations, have also been identified as health hazards in soil and groundwater.

Chronic exposure to benzene at sufficient concentrations is known to be associated with higher incidence of leukemia. Mercury and cyclodienes are known to induce higher incidences of kidney damage and some irreversible diseases. PCBs and cyclodienes are linked to liver toxicity. Organophosphates and carbomates can induce a chain of responses leading to neuromuscular blockage. Many chlorinated solvents induce liver changes, kidney changes and depression of the central nervous system. There is an entire spectrum of further health effects such as headache, nausea, fatigue, eye irritation and skin rash for the above cited and other chemicals. At sufficient dosages a large number of soil contaminants can cause death by exposure via direct contact, inhalation or ingestion of contaminants in groundwater contaminated through soil.

The Scottish Government has commissioned the Institute of Occupational Medicine to undertake a review of methods to assess risk to human health from contaminated land. The overall aim of the project is to work up guidance that should be useful to Scottish Local Authorities in assessing whether sites represent a significant possibility of significant harm (SPOSH) to human health. It is envisaged that the output of the project will be a short document providing high level guidance on health risk assessment with reference to existing published guidance and methodologies that have been identified as being particularly relevant and helpful. The project will examine how policy guidelines have been developed for determining the acceptability of risks to human health and propose an approach for assessing what constitutes unacceptable risk in line with the criteria for SPOSH as defined in the legislation and the Scottish Statutory Guidance.

Ecosystem Effects

Not unexpectedly, soil contaminants can have significant deleterious consequences for ecosystems. There are radical soil chemistry changes which can arise from the presence of many hazardous chemicals even at low concentration of the contaminant species. These changes can manifest in the alteration of metabolism of endemic microorganisms and arthropods resident in a given soil environment. The result can be virtual eradication of some of the primary food chain, which in turn could have major consequences for predator or consumer species. Even if the chemical effect on lower life forms is small, the lower pyramid levels of the food chain may ingest alien chemicals, which normally become more concentrated for each consuming rung of the food chain. Many of these effects are now well known, such as the concentration of persistent DDT materials for avian consumers, leading to weakening of egg shells, increased chick mortality and potential extinction of species.

Effects occur to agricultural lands which have certain types of soil contamination. Contaminants typically alter plant metabolism, often causing a reduction in crop yields. This has a secondary effect upon soil conservation, since the languishing crops cannot shield the Earth's soil from erosion. Some of these chemical contaminants have long half-lives and in other cases derivative chemicals are formed from decay of primary soil contaminants.

Cleanup Options

Cleanup or environmental remediation is analyzed by environmental scientists who utilize field measurement of soil chemicals and also apply computer models (GIS in Environmental Contamination) for analyzing transport and fate of soil chemicals. Various technologies have been developed for remediation of oil-contaminated soil/ sediments There are several principal strategies for remediation:

- Excavate soil and take it to a disposal site away from ready pathways for human or sensitive ecosystem contact. This technique also applies to dredging of bay muds containing toxins.
- Aeration of soils at the contaminated site (with attendant risk of creating air pollution)
- Thermal remediation by introduction of heat to raise subsurface temperatures sufficiently high to volatize chemical contaminants out of the soil for vapour extraction. Technologies include ISTD, electrical resistance heating (ERH), and ET-DSP^{tm}.
- Bioremediation, involving microbial digestion of certain organic chemicals. Techniques used in bioremediation include landfarming, biostimulation and bioaugmentating soil biota with commercially available microflora.
- Extraction of groundwater or soil vapor with an active electromechanical system, with subsequent stripping of the contaminants from the extract.
- Containment of the soil contaminants (such as by capping or paving over in place).
- Phytoremediation, or using plants (such as willow) to extract heavy metals
- Mycoremediation, or using fungus to metabolize contaminants and accumulate heavy metals.

By Country

Various national standards for concentrations of particular contaminants include the United States EPA Region 9 Preliminary Remediation Goals (U.S. PRGs), the U.S.

EPA Region 3 Risk Based Concentrations (U.S. EPA RBCs) and National Environment Protection Council of Australia Guideline on Investigation Levels in Soil and Groundwater.

People's Republic of China

The immense and sustained growth of the People's Republic of China since the 1970s has exacted a price from the land in increased soil pollution. The State Environmental Protection Administration believes it to be a threat to the environment, to food safety and to sustainable agriculture. According to a scientific sampling, 150 million mi (100,000 square kilometers) of China's cultivated land have been polluted, with contaminated water being used to irrigate a further 31.5 million mi (21,670 km^2.) and another 2 million mi (1,300 square kilometers) covered or destroyed by solid waste. In total, the area accounts for one-tenth of China's cultivatable land, and is mostly in economically developed areas. An estimated 12 million tonnes of grain are contaminated by heavy metals every year, causing direct losses of 20 billion yuan (US$2.57 billion).

United Kingdom

Generic guidance commonly used in the United Kingdom are the Soil Guideline Values published by the Department for Environment, Food and Rural Affairs (DEFRA) and the Environment Agency. These are screening values that demonstrate the minimal acceptable level of a substance. Above this there can be no assurances in terms of significant risk of harm to human health. These have been derived using the Contaminated Land Exposure Assessment Model (CLEA UK). Certain input parameters such as Health Criteria Values, age and land use are fed into CLEA UK to obtain a probabilistic output.

Guidance by the Inter Departmental Committee for the Redevelopment of Contaminated Land (ICRCL) has been formally withdrawn by DEFRA, for use as a prescriptive document to determine the potential need for remediation or further assessment.

The CLEA model published by DEFRA and the Environment Agency (EA) in March 2002 sets a framework for the appropriate assessment of risks to human health from contaminated land, as required by Part IIA of the Environmental Protection Act 1990. As part of this framework, generic Soil Guideline Values (SGVs) have currently been derived for ten contaminants to be used as "intervention values". These values should not be considered as remedial targets but values above which further detailed assessment should be considered.

Three sets of CLEA SGVs have been produced for three different land uses, namely

- residential (with and without plant uptake)
- allotments
- commercial/industrial

It is intended that the SGVs replace the former ICRCL values. It should be noted that the CLEA SGVs relate to assessing chronic (long term) risks to human health and do not apply to the protection of ground workers during construction, or other potential receptors such as groundwater, buildings, plants or other ecosystems. The CLEA SGVs are not directly applicable to a site completely covered in hardstanding, as there is no direct exposure route to contaminated soils.

To date, the first ten of fifty-five contaminant SGVs have been published, for the following: arsenic, cadmium, chromium, lead, inorganic mercury, nickel, selenium ethyl benzene, phenol and toluene. Draft SGVs for benzene, naphthalene and xylene have been produced but their publication is on hold. Toxicological data (Tox) has been published for each of these contaminants as well as for benzo[a]pyrene, benzene, dioxins, furans and dioxin-like PCBs, naphthalene, vinyl chloride, 1,1,2,2 tetrachloroethane and 1,1,1,2 tetrachloroethane, 1,1,1 trichloroethane, tetrachloroethene, carbon tetrachloride, 1,2-dichloroethane, trichloroethene and xylene. The SGVs for ethyl benzene, phenol and toluene are dependent on the soil organic matter (SOM) content (which can be calculated from the total organic carbon (TOC) content). As an initial screen the SGVs for 1% SOM are considered to be appropriate.

Canada

India

In March 2009, the issue of Uranium poisoning in Punjab attracted press coverage. It was alleged to be caused by fly ash ponds of thermal power stations, which reportedly lead to severe birth defects in children in the Faridkot and Bhatinda districts of Punjab. The news reports claimed the uranium levels were more than 60 times the maximum safe limit. In 2012, the Government of India confirmed that the ground water in Malwa belt of Punjab has uranium metal that is 50% above the trace limits set by the United Nations' World Health Organization. Scientific studies, based on over 1000 samples from various sampling points, could not trace the source to fly ash and any sources from thermal power plants or industry as originally alleged. The study also revealed that the uranium concentration in ground water of Malwa district is not 60 times the WHO limits, but only 50% above the WHO limit in 3 locations. This highest concentration found in samples was less than those found naturally in ground waters currently used for human purposes elsewhere, such as Finland. Research is underway to identify natural or other sources for the uranium.

Land Degradation

Serious land degradation in Nauru after the depletion of the phosphate cover through mining

Land degradation is a process in which the value of the biophysical environment is affected by a combination of human-induced processes acting upon the land. It is viewed as any change or disturbance to the land perceived to be deleterious or undesirable. Natural hazards are excluded as a cause; however human activities can indirectly affect phenomena such as floods and bush fires.

This is considered to be an important topic of the 21st century due to the implications land degradation has upon agronomic productivity, the environment, and its effects on food security. It is estimated that up to 40% of the world's agricultural land is seriously degraded.

Measures

Land degradation is a broad term that can be applied differently across a wide range of scenarios. There are four main ways of looking at land degradation and its impact on the environment around it:

1. A temporary or permanent decline in the productive capacity of the land. This can be seen through a loss of biomass, a loss of actual productivity or in potential productivity, or a loss or change in vegetative cover and soil nutrients.

2. Action in the land's capacity to provide resources for human livelihoods. This can be measured from a base line of past land use.

3. Loss of biodiversity: A loss of range of species or ecosystem complexity as a decline in the environmental quality.

4. Shifting ecological risk: increased vulnerability of the environment or people to destruction or crisis. This is measured through a base line in the form of pre-existing risk of crisis or destruction.

A problem with defining land degradation is that what one group of people might view as degradation, others might view as a benefit or opportunity. For example, planting crops at a location with heavy rainfall and steep slopes would create scientific and environmental concern regarding the risk of soil erosion by water, yet farmers could view the location as a favourable one for high crop yields.

Different Types

In addition to the usual types of land degradation that have been known for centuries (water, wind and mechanical erosion, physical, chemical and biological degradation), four other types have emerged in the last 50 years:

- pollution, often chemical, due to agricultural, industrial, mining or commercial activities;
- loss of arable land due to urban construction;
- artificial radioactivity, sometimes accidental;
- land-use constraints associated with armed conflicts.

Overall, 36 types of land degradation can be assessed. All are induced or aggravated by human activities, e.g. sheet erosion, silting, aridification, salinization, urbanization, etc.

Causes

Overgrazing by livestock can lead to land degradation

Land degradation is a global problem largely related to agricultural use. The major causes include:

- Land clearance, such as clearcutting and deforestation
- Agricultural depletion of soil nutrients through poor farming practices
- Livestock including overgrazing and overdrafting
- Inappropriate irrigation and overdrafting
- Urban sprawl and commercial development
- Vehicle off-roading
- Quarrying of stone, sand, ore and minerals
- Increase in field size due to economies of scale, reducing shelter for wildlife, as hedgerows and copses disappear
- Exposure of naked soil after harvesting by heavy equipment
- Monoculture, destabilizing the local ecosystem
- Dumping of non-biodegradable trash, such as plastics
- Soil degradation, e.g.
    - Soil contamination
    - Soil erosion
    - Soil acidification
    - Loss of soil carbon

Effects

Overcutting of vegetation occurs when people cut forests, woodlands and shrublands—to obtain timber, fuelwood and other products—at a pace exceeding the rate of natural regrowth. This is frequent in semi-arid environments, where fuelwood shortages are often severe.

Overgrazing is the grazing of natural pastures at stocking intensities above the livestock carrying capacity; the resulting decrease in the vegetation cover is a leading cause of wind and water erosion. It is a significant factor in Afghanistan. The growing population pressure, during 1980-1990, has led to decreases in the already small areas of agricultural land per person in six out of eight countries (14% for India and 22% for Pakistan).

Soil erosion in a wheat field near Pullman, USA.

Population pressure also operates through other mechanisms. Improper agricultural practices, for instance, occur only under constraints such as the saturation of good lands under population pressure which leads settlers to cultivate too shallow or too steep soils, plough fallow land before it has recovered its fertility, or attempt to obtain multiple crops by irrigating unsuitable soils.

High population density is not always related to land degradation. Rather, it is the practices of the human population that can cause a landscape to become degraded. Populations can be a benefit to the land and make it more productive than it is in its natural state. Land degradation is an important factor of internal displacement in many African and Asian countries

Severe land degradation affects a significant portion of the Earth's arable lands, decreasing the wealth and economic development of nations. As the land resource base becomes less productive, food security is compromised and competition for dwindling resources increases, the seeds of famine and potential conflict are sown.

Sensitivity and Resilience

Sensitivity and resilience are measures of the vulnerability of a landscape to degradation. These two factors combine to explain the degree of vulnerability. Sensitivity is the degree to which a land system undergoes change due to natural forces, human intervention or a combination of both. Resilience is the ability of a landscape to absorb change, without significantly altering the relationship between the relative importance and numbers of individuals and species that compose the community. It also refers to the ability of the region to return to its original state after being changed in some way. The resilience of a landscape can be increased or decreased through human interaction based upon different methods of land-use management. Land that is degraded becomes less resilient than undegraded land, which can lead to even further degradation through shocks to the landscape.

Climate Change

Significant land degradation from seawater inundation, particularly in river deltas and on low-lying islands, is a potential hazard that was identified in a 2007 IPCC report.

IPCC
Assessment reports:
First (1990)
1992 sup.
Second (1995)
Third (2001)
Fourth (2007)
Fifth (2014)
UNFCCC | WMO | UNEP

As a result of sea-level rise from climate change, salinity levels can reach levels where agriculture becomes impossible in very low-lying areas.

Soil Retrogression and Degradation

Soil retrogression and degradation are two regressive evolution processes associated with the loss of equilibrium of a stable soil. Retrogression is primarily due to soil erosion and corresponds to a phenomenon where succession reverts the land to its natural physical state. Degradation is an evolution, different from natural evolution, related to the local climate and vegetation. It is due to the replacement of primary plant communities (known as climax vegetation) by the secondary communities. This replacement modifies the humus composition and amount, and affects the formation of the soil. It is directly related to human activity. Soil degradation may also be viewed as any change or ecological disturbance to the soil perceived to be deleterious or undesirable.

General

At the beginning of soil formation, the bare rock out crops is gradually colonized by pioneer species (lichens and mosses). They are succeeded by herbaceous vegetation, shrubs and finally forest. In parallel, the first humus-bearing horizon is formed (the A horizon), followed by some mineral horizons (B horizons). Each successive stage is characterized by a certain association of soil/vegetation and environment, which defines an ecosystem.

Intensive tillage result on soil degradation.

Willow hedge strengthened with fascines for the limitation of runoff, north of France.

After a certain time of parallel evolution between the ground and the vegetation, a state of steady balance is reached. This stage of development is called climax by some ecologists and "natural potential" by others. Succession is the evolution towards climax. Regardless of its name, the equilibrium stage of primary succession is the highest natural form of development that the environmental factors are capable of producing.

The cycles of evolution of soils have very variable durations, between tens, hundreds, or thousands of years for quickly evolving soils (A horizon only) to more than a million years for slowly developing soils. The same soil may achieve several successive steady state conditions during its existence, as exhibited by the Pygmy forest sequence in Mendocino County, California. Soils naturally reach a state of high productivity, from which they naturally degrade as mineral nutrients are removed from the soil system. Thus older soils are more vulnerable to the effects of induced retrogression and degradation.

Ecological Factors Influencing Soil Formation

There are two types of ecological factors influencing the evolution of a soil (through alteration and humification). These two factors are extremely significant to explain the evolution of soils of short development.

- A first type of factor is the average climate of an area and the vegetation which is associated (biome).
- A second type of factor is more local, and is related to the original rock and local drainage. This type of factor explains appearance of specialized associations (ex peat bogs).

Biorhexistasy Theory

The destruction of the vegetation implies the destruction of evoluted soils, or a regressive evolution. Cycles of succession-regression of soils follow one another within short intervals of time (human actions) or long intervals of time (climate variations).

The climate role in the deterioration of the rocks and the formation of soils lead to the formulation of the theory of the biorhexistasy.

- In wet climate, the conditions are favorable to the deterioration of the rocks (mostly chemically), the development of the vegetation and the formation of soils; this period favorable to life is called biostasy.
- In dry climate, the rocks exposed are mostly subjected to mechanical disintegration which produces coarse detrital materials: this is referred to as rhexistasy.

Perturbations of the Balance of a Soil

When the state of balance, characterized by the ecosystem climax is reached, it tends to be maintained stable in the course of time. The vegetation installed on the ground provides the humus and ensures the ascending circulation of the matters. It protects the ground from erosion by playing the role of barrier (for example, protection from water and wind). Plants can also reduce erosion by binding the particles of the ground to their roots.

A disturbance of climax will cause retrogression, but often, secondary succession will start to guide the evolution of the system after that disturbance. Secondary succession is much faster than primary because the soil is already formed, although deteriorated and needing restoration as well.

However, when a significant destruction of the vegetation takes place (of natural origin such as an avalanche or human origin), the disturbance undergone by the ecosystem is too important. In this latter case, erosion is responsible for the destruction of the upper horizons of the ground, and is at the origin of a phenomenon of reversion to pioneer conditions. The phenomenon is called retrogression and can be partial or total (in this case, nothing remains beside bare rock). For example, the clearing of an inclined ground, subjected to violent rains, can lead to the complete destruction of the soil. Man can deeply modify the evolution of the soils by direct and brutal action, such as clearing, abusive cuts, forest pasture, litters raking. The climax vegetation is gradually replaced and the soil modified (example: replacement of leafy tree forests by moors or pines plantations). Retrogression is often related to very old human practices.

Influence of Human Activity

Soil erosion is the main factor for soil degradation and is due to several mechanisms: water erosion, wind erosion, chemical degradation and physical degradation.

Erosion is strongly related to human activity. For example, roads which increase impermeable surfaces lead to streaming and ground loss. Agriculture also accelerates soil erosion (increase of field size, correlated to hedges and ditches removal). Meadows are in regression to the profit of plowed lands. Spring cultures (sunflower, corn, beet) surfaces are increasing and leave the ground naked in winter. Sloping grounds are gradually colonized by vine. Lastly, use of herbicides leaves the ground naked between each crop. New cultural practices, such as mechanization also increases the risks of erosion. Fertilization by mineral manures rather than organic manure gradually destructure the soil. Many scientistsobserved a gradual decrease of soil organic matter content in soils, as well as a decrease of soil biological activity (in particular, in relation to chemical uses). Lastly, deforestation, in particular, is responsible for degradation of forest soils.

Agriculture increases the risk of erosion through its disturbance of vegetation by way of:

- Overgrazing of animals
- Monoculture planting
- Row cropping
- Tilling or plowing
- Crop removal
- Land-use conversion

Consequences of Soil Regression and Degradation

- yields impact: Recent increases in the human population have placed a great strain on the world's soil systems. More than 6 billion people are now using about 38% of the land area of the Earth to raise crops and livestock. Many soils suffer from various types of degradation, that can ultimately reduce their ability to produce food resources. Slight degradation refers to land where yield potential has been reduced by 10%, moderate degradation refers to a yield decrease from 10-50%. Severely degraded soils have lost more than 50% of their potential. Most severely degraded soils are located in developing countries.

- natural disasters: natural disasters such as mud flows, floods are responsible for the death of many living beings each year.
- deterioration of the water quality: the increase in the turbidity of water and the contribution of nitrogen and of phosphorus can result in eutrophication. Soils particles in surface waters are also accompanied by agricultural inputs and by some pollutants of industrial, urban and road origin (such as heavy metals). The ecological impact of agricultural inputs (such as weed killer) is known but difficult to evaluate because of the multiplicity of the products and their broad spectrum of action.
- biological diversity: soil degradation may involve perturbation of microbial communities, disappearance of the climax vegetation and decrease in animal habitat, thus leading to a biodiversity loss and animal extinction.
- economic loss: the estimated costs for land degradation are US $40 billion per year. This figure does not take into account negative externalities such as the increased use of fertilizers, loss of biodiversity and loss of unique landscapes.

Soil Enhancement, Rebuilding and Regeneration

Problems of soil erosion can be fought, and certain practices can lead to soil enhancement and rebuilding. Even though simple, methods for reducing erosion are often not chosen because these practices outweigh the short-term benefits. Rebuilding is especially possible through the improvement of soil structure, addition of organic matter and limitation of runoff. However, these techniques will never totally succeed to restore a soil (and the fauna and flora associated to it) that took more than 1000 years to build up. Soil regeneration is the reformation of degraded soil through biological, chemical, and or physical processes.

When productivity declined in the low-clay soils of northern Thailand, farmers initially responded by adding organic matter from termite mounds, but this was unsustainable in the long-term. Scientists experimented with adding bentonite, one of the smectite family of clays, to the soil. In field trials, conducted by scientists from the International Water Management Institute in cooperation with Khon Kaen University and local farmers, this had the effect of helping retain water and nutrients. Supplementing the farmer's usual practice with a single application of 200 kg bentonite per rai (6.26 rai =

1 hectare) resulted in an average yield increase of 73%. More work showed that applying bentonite to degraded sandy soils reduced the risk of crop failure during drought years.

In 2008, three years after the initial trials, IWMI scientists conducted a survey among 250 farmers in northeast Thailand, half who had applying bentonite to their fields and half who had not. The average output for those using the clay addition was 18% higher than for non-clay users. Using the clay had enabled some farmers to switch to growing vegetables, which need more fertile soil. This helped to increase their income. The researchers estimated that 200 farmers in northeast Thailand and 400 in Cambodia had adopted the use of clays, and that a further 20,000 farmers were introduced to the new technique.

References

- Blanco-Canqui, Humberto; Rattan, Lal (2008). "Soil and water conservation". Principles of soil conservation and management. Dordrecht: Springer. pp. 1–20. ISBN 9781402087097.
- Toy, Terrence J.; Foster, George R.; Renard, Kenneth G. (2002). Soil erosion : processes, prediction, measurement, and control. New York, NY: Wiley. ISBN 9780471383697.
- Food and Agriculture Organization (1965). "Types of erosion damage". Soil Erosion by Water: Some Measures for Its Control on Cultivated Lands. United Nations. pp. 23–25. ISBN 978-92-5-100474-6.
- Borah, Deva K. et al. (2008). "Watershed sediment yield". In Garcia, Marcelo H. Sedimentation Engineering: Processes, Measurements, Modeling, and Practice. ASCE Publishing. p. 828. ISBN 978-0-7844-0814-8.
- Nancy D. Gordon (2004-06-01). "Erosion and Scour". Stream hydrology: an introduction for ecologists. ISBN 978-0-470-84357-4.
- Zheng, Xiaojing & Huang, Ning (2009). Mechanics of Wind-Blown Sand Movements. Springer. pp. 7–8. ISBN 978-3-540-88253-4.
- Cornelis, Wim S. (2006). "Hydroclimatology of wind erosion in arid and semi-arid environments". In D'Odorico, Paolo & Porporato, Amilcare. Dryland Ecohydrology. Springer. p. 141. ISBN 978-1-4020-4261-4.
- Blanco-Canqui, Humberto; Rattan, Lal (2008). "Wind erosion". Principles of soil conservation and management. Dordrecht: Springer. pp. 54–80. ISBN 9781402087097.
- Balba, A. Monem (1995). "Desertification: Wind erosion". Management of Problem Soils in Arid Ecosystems. CRC Press. p. 214. ISBN 978-0-87371-811-0.
- Wiggs, Giles F.S. (2011). "Geomorphological hazards in drylands". In Thomas, David S.G. Arid Zone Geomorphology: Process, Form and Change in Drylands. John Wiley & Sons. p. 588. ISBN 978-0-470-71076-0.
- Van Beek, Rens (2008). "Hillside processes: mass wasting, slope stability, and erosion". In Norris, Joanne E. et al. Slope Stability and Erosion Control: Ecotechnological Solutions. Springer. ISBN 978-1-4020-6675-7.
- Blanco-Canqui, Humberto; Rattan, Lal (2008). "Water erosion". Principles of soil conservation and management. Dordrecht: Springer. pp. 21–53. ISBN 9781402087097.

12

Integrated Areas of Soil Science

The following chapter will emphasize on the different scientific aspects of soil. The chapter has elaborated and discussed at length, multiple integrated areas of study like soil physics, soil chemistry and soil biology which are closely associated with the umbrella field of soil science.

Soil Physics

Soil physics is the study of soil physical properties and processes. It is applied to management and prediction under natural and managed ecosystems. Soil physics deals with the dynamics of physical soil components and their phases as solid, liquids, and gases. It draws on the principles of physics, physical chemistry, engineering, and meteorology. It is especially important in this day and age because most farmers require an understanding of agroecosystems. Soil physics applies these principles to address practical problems of agriculture, ecology, and engineering.

Prominent Soil Physicists

- Edgar Buckingham (1867–1940)

  The theory of gas diffusion in soil and vadose zone water flow in soil.

- Lorenzo A. Richards (1904–1993)

  General transport of water in unsaturated soil, measurement of soil water potential using tensiometer.

- John R. Philip (1927–1999)

  Analytical solution to general soil water transport, Environmental Mechanics.

Soil Chemistry

Soil chemistry is the study of the chemical characteristics of soil. Soil chemistry is affected by mineral composition, organic matter and environmental factors.

Soil chemistry is the study of the chemical characteristics of soil. Soil chemistry is affected by mineral composition, organic matter and environmental factors. Soil - (i) The unconsolidated mineral or organic material on the immediate surface of the Earth that serves as a natural medium for the growth of land plants. (ii) The unconsolidated mineral or organic matter on the surface of the Earth that has been subjected to and shows effects of genetic and environmental factors of: climate (including water and temperature effects), and macro- and micro organisms, conditioned by relief, acting on parent material over a period of time. A product-soil differs from the material from which it is derived in many physical, chemical, biological, and morphological properties and characteristics. This definition is from Soil Taxonomy, second edition.

Key facts you should know about Soil:

Soil makes up the outermost layer of our planet and is formed from rocks and decaying plants and Animals. Soil has varying amounts of organic matter (living and dead organisms), minerals, and nutrients. • An average soil sample is 45 percent minerals, 25 percent water, 25 percent air, and five percent organic matter. Different-sized mineral particles, such as sand, silt, and clay, give soil its texture. • Topsoil is the most productive soil layer. • Ten tonnes of topsoil spread evenly over a hectare is only as thick as a one Euro coin. • Natural processes can take more than 500 years to form 2 centimeters of topsoil. • In some cases, 5 tonnes of animal life can live in one hectare of soil. • Fungi and bacteria help break down organic matter in the soil. • Earthworms digest organic matter, recycle nutrients, and make the surface soil richer. • Roots loosen the soil, allowing oxygen to penetrate. This benefits animals living in the soil. They also hold soil together and help prevent erosion. • A fully functioning soil reduces the risk of floods and protects underground water supplies by neutralizing or filtering out potential pollutants and storing as much as 3750 tonnes of water per hectare. • Scientists have found 10,000 types of soil in Europe and about 70,000 types of soil in the United States. • Soil stores 10% of the world's carbon dioxide emissions.

Environmental Soil Chemistry

A knowledge of environmental soil chemistry is paramount to predicting the fate of contaminants, as well as the processes by which they are initially released into the soil. Once a chemical is exposed to the soil environment myriad chemical reactions can occur that may increase or decrease contaminant toxicity. These reactions include adsorption/desorption, precipitation, polymerization, dissolution, complexation and oxidation/reduction. These reactions are often disregarded by scientists and engineers involved with environmental remediation. Understanding these processes enable us to better predict the fate and toxicity of contaminants and provide the knowledge to develop scientifically correct, and cost-effective remediation strategies.

Concepts

- Anion and cation exchange capacity
- Soil pH
- Mineral formation and transformation processes
- Clay mineralogy
- Sorption and precipitation reactions in soil
- Oxidation-reduction reactions
- Chemistry of problem soils

Soil Biology

Soil biology is the study of microbial and faunal activity and ecology in soil. Soil life, soil biota, soil fauna, or edaphon is a collective term that encompasses all the organisms that spend a significant portion of their life cycle within a soil profile, or at the soil-litter interface. These organisms include earthworms, nematodes, protozoa, fungi, bacteria and different arthropods. Soil biology plays a vital role in determining many soil characteristics. The decomposition of organic matter by soil organisms has an immense influence on soil fertility, plant growth, soil structure, and carbon storage. As a relatively new science, much remains unknown about soil biology and their effects on soil ecosystems.

Overview

The soil is home to a large proportion of the world's biodiversity. The links between soil organisms and soil functions are observed to be incredibly complex. The interconnectedness and complexity of this soil 'food web' means any appraisal of soil function must necessarily take into account interactions with the living communities that exist within the soil. We know that soil organisms break down organic matter, making nutrients available for uptake by plants and other organisms. The nutrients stored in the bodies of soil organisms prevent nutrient loss by leaching. Microbial exudates act to maintain soil structure, and earthworms are important in bioturbation. However, we find that we don't understand critical aspects about how these populations function and interact. The discovery of glomalin in 1995 indicates that we lack the knowledge to correctly answer some of the most basic questions about the biogeochemical cycle in soils. We have much work ahead to gain a better understanding of how soil biological components affect us and the biosphere

In balanced soil, plants grow in an active and steady environment. The mineral content of the soil and its heartiful structure are important for their well-being, but it is the life in the earth that powers its cycles and provides its fertility. Without the activities of soil organisms, organic materials would accumulate and litter the soil surface, and there would be no food for plants. The soil biota includes:

- Megafauna: size range - 20 mm upward, e.g. moles, rabbits, and rodents.
- macrofauna: size range - 2 to 20 mm, e.g. woodlice, earthworms, beetles, centipedes, slugs, snails, ants, and harvestmen.
- Mesofauna: size range - 100 micrometres to 2 mm, e.g. tardigrades, mites and springtails.
- Microfauna and Microflora: size range - 1 to 100 micrometres, e.g. yeasts, bacteria (commonly actinobacteria), fungi, protozoa, roundworms, and rotifers.

Of these, bacteria and fungi play key roles in maintaining a healthy soil. They act as decomposers that break down organic materials to produce detritus and other breakdown products. Soil detritivores, like earthworms, ingest detritus and decompose it. Saprotrophs, well represented by fungi and bacteria, extract soluble nutrients from delitro. The ants (macrofaunas) help by breaking down in the same way but they also provide the motion part as they move in their armies. Also the rodents, wood-eaters help the soil to be more absorbent.

Soil Life Table

Scope

Soil biology involves work in the following areas:

- Modelling of biological processes and population dynamics
- Soil biology, physics and chemistry: occurrence of physicochemical parameters and surface properties on biological processes and population behavior
- Population biology and molecular ecology: methodological development and contribution to study microbial and faunal populations; diversity and population dynamics; genetic transfers, influence of environmental factors
- Community ecology and functioning processes: interactions between organisms and mineral or organic compounds; involvement of such interactions in soil pathogenicity; transformation of mineral and organic compounds, cycling of elements; soil structuration

Complementary disciplinary approaches are necessarily utilized which involve molecular biology, genetics, ecophysiology, biogeography, ecology, soil processes, organic matter, nutrient dynamics and landscape ecology.

Bacteria

Bacteria are single-cell organisms and the most numerous denizens of agriculture, with populations ranging from 100 million to 3 billion in a gram. They are capable of very rapid reproduction by binary fission (dividing into two) in favourable conditions. One bacterium is capable of producing 16 million more in just 24 hours. Most soil bacteria live close to plant roots and are often referred to as rhizobacteria. Bacteria live in soil water, including the film of moisture surrounding soil particles, and some are able to swim by means of flagella. The majority of the beneficial soil-dwelling bacteria need oxygen (and are thus termed aerobic bacteria), whilst those that do not require air are referred to as anaerobic, and tend to cause putrefaction of dead organic matter. Aerobic bacteria are most active in a soil that is moist (but not saturated, as this will deprive aerobic bacteria of the air that they require), and neutral soil pH, and where there is plenty of food (carbohydrates and micronutrients from organic matter) available. Hostile conditions will not completely kill bacteria; rather, the bacteria will stop growing and get into a dormant stage, and those individuals with pro-adaptive mutations may compete better in the new conditions. Some gram-positive bacteria produce spores in order to wait for more favourable circumstances, and gram-negative bacteria get into a "nonculturable" stage. Bacteria are colonized by persistent viral agents (bacteriophages) that determine gene word order in bacterial host.

From the organic gardener's point of view, the important roles that bacteria play are:

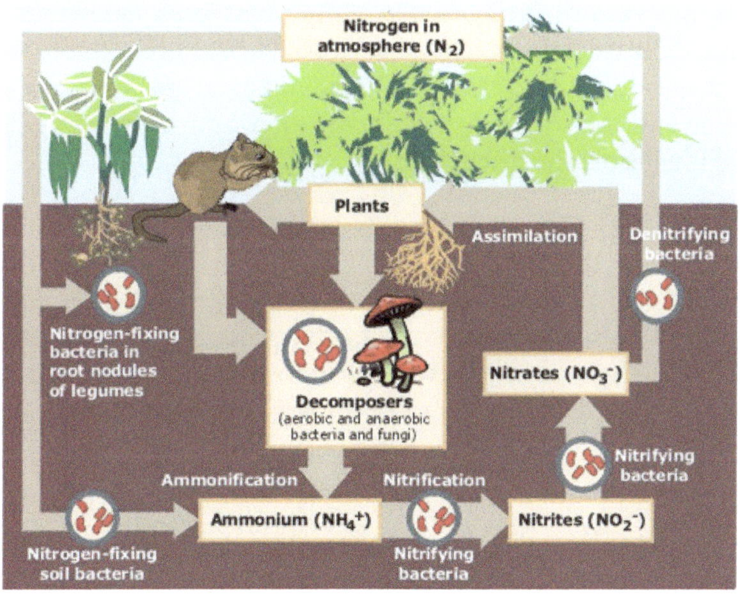

The nitrogen cycle (Turkish)

Nitrification

Nitrification is a vital part of the nitrogen cycle, wherein certain bacteria (which manufacture their own carbohydrate supply without using the process of photosynthesis) are able to transform nitrogen in the form of ammonium, which is produced by the decomposition of proteins, into nitrates, which are available to growing plants, and once again converted to proteins.

Nitrogen Fixation

In another part of the cycle, the process of nitrogen fixation constantly puts additional nitrogen into biological circulation. This is carried out by free-living nitrogen-fixing bacteria in the soil or water such as *Azotobacter*, or by those that live in close symbiosis with leguminous plants, such as rhizobia. These bacteria form colonies in nodules they create on the roots of peas, beans, and related species. These are able to convert nitrogen from the atmosphere into nitrogen-containing organic substances.

Denitrification

While nitrogen fixation converts nitrogen from the atmosphere into organic compounds, a series of processes called denitrification returns an approximately equal amount of nitrogen to the atmosphere. Denitrifying bacteria tend to be anaerobes, or facultatively anaerobes (can alter between the oxygen dependent and oxygen independent types of metabolisms), including *Achromobacter* and *Pseudomonas*. The purification process caused by oxygen-free conditions converts nitrates and nitrites in soil into nitrogen gas or into gaseous compounds such as nitrous oxide or nitric oxide. In excess, denitrification can lead to overall losses of available soil nitrogen and subsequent loss of soil fertility. However, fixed nitrogen may circulate many times between organisms and the soil before denitrification returns it to the atmosphere. The diagram above illustrates the nitrogen cycle.

Actinobacteria

Actinobacteria are critical in the decomposition of organic matter and in humus formation, and their presence is responsible for the sweet "earthy" aroma associated with a good healthy soil. They require plenty of air and a pH between 6.0 and 7.5, but are more tolerant of dry conditions than most other bacteria and fungi.

Fungi

A gram of garden soil can contain around one million fungi, such as yeasts and moulds. Fungi have no chlorophyll, and are not able to photosynthesise. They cannot use atmospheric carbon dioxide as a source of carbon, therefore they are chemo-heterotrophic, meaning that, like animals, they require a chemical source of energy rather than being able to use light as an energy source, as well as organic substrates to get carbon for growth and development.

Many fungi are parasitic, often causing disease to their living host plant, although some have beneficial relationships with living plants, as illustrated below. In terms of soil and humus creation, the most important fungi tend to be saprotrophic; that is, they live on dead or decaying organic matter, thus breaking it down and converting it to forms that are available to the higher plants. A succession of fungi species will colonise the dead matter, beginning with those that use sugars and starches, which are succeeded by those that are able to break down cellulose and lignins.

Fungi spread underground by sending long thin threads known as mycelium throughout the soil; these threads can be observed throughout many soils and compost heaps. From the mycelia the fungi is able to throw up its fruiting bodies, the visible part above the soil (e.g., mushrooms, toadstools, and puffballs), which may contain millions of spores. When the fruiting body bursts, these spores are dispersed through the air to settle in fresh environments, and are able to lie dormant for up to years until the right conditions for their activation arise or the right food is made available.

Mycorrhizae

Those fungi that are able to live symbiotically with living plants, creating a relationship that is beneficial to both, are known as Mycorrhizae (from *myco* meaning fungal and *rhiza* meaning root). Plant root hairs are invaded by the mycelia of the mycorrhiza, which lives partly in the soil and partly in the root, and may either cover the length of the root hair as a sheath or be concentrated around its tip. The mycorrhiza obtains the carbohydrates that it requires from the root, in return providing the plant with nutrients including nitrogen and moisture. Later the plant roots will also absorb the mycelium into its own tissues.

Beneficial mycorrhizal associations are to be found in many of our edible and flowering crops. Shewell Cooper suggests that these include at least 80% of the *brassica* and *solanum* families (including tomatoes and potatoes), as well as the majority of tree species, especially in forest and woodlands. Here the mycorrhizae create a fine underground mesh that extends greatly beyond the limits of the tree's roots, greatly increasing their feeding range and actually causing neighbouring trees to become physically interconnected. The benefits of mycorrhizal relations to their plant partners are not limited to nutrients, but can be essential for plant reproduction: In situations where little light is able to reach the forest floor, such as the North American pine forests, a young seedling cannot obtain sufficient light to photosynthesise for itself and will not grow properly in a sterile soil. But, if the ground is underlain by a mycorrhizal mat, then the developing seedling will throw down roots that can link with the fungal threads and through them obtain the nutrients it needs, often indirectly obtained from its parents or neighbouring trees.

David Attenborough points out the plant, fungi, animal relationship that creates a "Three way harmonious trio" to be found in forest ecosystems, wherein the plant/

fungi symbiosis is enhanced by animals such as the wild boar, deer, mice, or flying squirrel, which feed upon the fungi's fruiting bodies, including truffles, and cause their further spread (*Private Life Of Plants*, 1995). A greater understanding of the complex relationships that pervade natural systems is one of the major justifications of the organic gardener, in refraining from the use of artificial chemicals and the damage these might cause.

Recent research has shown that arbuscular mycorrhizal fungi produce glomalin, a protein that binds soil particles and stores both carbon and nitrogen. These glomalin-related soil proteins are an important part of soil organic matter.

Earthworms Ants and Termites

Earthworms, ants and termites mix the soil as they burrow, significantly affecting soil formation. Earthworms ingest soil particles and organic residues, enhancing the availability of plant nutrients in the material that passes through and out their bodies. By aerating and stirring the soil, and by increasing the stability of soil aggregates, these organisms help assure the ready infiltration of water.

References

- Lal, Rattan; Manoj Shukla (2004). Principles of Soil Physics. CRC Press. p. 5. ISBN 0-8247-5324-0.
- Encyclopedia of Soil Science, edts. Ward Chesworth, 2008, Uniw. of Guelph Canada, Publ. Springer, ISBN 978-1-4020-3994-2

13

Tools and Techniques of Soil Science

The tools and techniques employed to study soil are of great importance. The following chapter will provide a comprehensive insight into the major tools and techniques that are used in the subject of soil science such as soil nailing, pedometrics, soil compaction, etc. They have been discussed elaborately so as to give the readers a complete understanding of soil science.

Soil Moisture Sensor

A simple soil moisture sensor for gardeners.

Soil moisture sensors measure the volumetric water content in soil. Since the direct gravimetric measurement of free soil moisture requires removing, drying, and weighting of a sample, soil moisture sensors measure the volumetric water content indirectly by using some other property of the soil, such as electrical resistance, dielectric constant, or interaction with neutrons, as a proxy for the moisture content. The relation between the measured property and soil moisture must be calibrated and may vary depending on environmental factors such as soil type, temperature, or electric conductivity. Reflected microwave radiation is affected by the soil moisture and is used for remote sensing in hydrology and agriculture. Portable probe instruments can be used by farmers or gardeners.

Soil moisture sensors typically refer to sensors that estimate volumetric water content. Another class of sensors measure another property of moisture in soils called water potential; these sensors are usually referred to as soil water potential sensors and include tensiometers and gypsum blocks.

Technology

Technologies commonly used to indirectly measure volumetric water content (soil moisture) include)

- Frequency Domain Reflectometry (FDR): The dielectric constant of a certain volume element around the sensor is obtained by measuring the operating frequency of an oscillating circuit.
- Time Domain Transmission (TDT) and Time Domain Reflectometry (TDR): The dielectric constant of a certain volume element around the sensor is obtained by measuring the speed of propagation along a buried transmission line.
- Neutron moisture gauges: The moderator properties of water for neutrons are utilized to estimate soil moisture content between a source and detector probe.
- Soil resistivity: Measuring how strongly the soil resists the flow of electricity between two electrodes can be used to determine the soil moisture content.
- Galvanic cell: The amount of water present can be determined based on the voltage the soil produces because water acts as an electrolyte and produces electricity. The technology behind this concept is the galvanic cell.

Application

Agriculture

Measuring soil moisture is important for agricultural applications to help farmers manage their irrigation systems more efficiently. Knowing the exact soil moisture conditions on their fields, not only are farmers able to generally use less water to grow a crop, they are also able to increase yields and the quality of the crop by improved management of soil moisture during critical plant growth stages.

Landscape Irrigation

In urban and suburban areas, landscapes and residential lawns are using soil moisture sensors to interface with an irrigation controller. Connecting a soil moisture sensor to a simple irrigation clock will convert it into a "smart" irrigation controller that prevents irrigation cycles when the soil is already wet, e.g. following a recent rainfall event.

Golf courses are using soil moisture sensors to increase the efficiency of their irrigation systems to prevent over-watering and leaching of fertilizers and other chemicals into the ground.

Research

Soil moisture sensors are used in numerous research applications, e.g. in agricultural science and horticulture including irrigation planning, climate research, or environmental science including solute transport studies and as auxiliary sensors for soil respiration measurements.

Simple Sensors for Gardeners

Relatively cheap and simple devices that do not require a power source are available for checking whether plants have sufficient moisture to thrive. After inserting a probe into the soil for approximately 60 seconds a meter indicates if the soil is too dry, moist or wet for plants.

Soil Nailing

Soil nailing is a construction technique that can be used as a remedial measure to treat unstable natural soil slopes or as a construction technique that allows the safe over-steepening of new or existing soil slopes. The technique involves the insertion of relatively slender reinforcing elements into the slope – often general purpose reinforcing bars (rebar) although proprietary solid or hollow-system bars are also available. Solid bars are usually installed into pre-drilled holes and then grouted into place using a separate grout line, whereas hollow bars may be drilled and grouted simultaneously by the use of a sacrificial drill bit and by pumping grout down the hollow bar as drilling progresses. Kinetic methods of firing relatively short bars into soil slopes have also been developed. Bars installed using drilling techniques are usually fully grouted and installed at a slight downward inclination with bars installed at regularly spaced points across the slope face. A rigid facing (often pneumatically applied concrete, otherwise known as shotcrete) or isolated soil nail head plates may be used at the surface. Alternatively a flexible reinforcing mesh may be held against the soil face beneath the head plates. Rabbit proof wire mesh and environmental erosion control fabrics and may be used in conjunction with flexible mesh facing where environmental conditions dictate.

Soil nail components may also be used to stabilize retaining walls or existing fill slopes (embankments and levees); this is normally undertaken as a remedial measure.

Since its first application using modern techniques in Versailles in 1972, soil nailing is now a well-established technique around the world. One of the first national guideline publications for soil nailing was produced in Japan in 1987. The U.S. Federal Highway Administration issued guideline publications in 1996 and 2003.

Preliminary Analysis

Four main points to be considered in determining if soil nailing would be an effective retention technique are as follows. First, the existing ground conditions should be examined. Next, the advantages and disadvantages for a soil nail wall should be assessed for the particular application being considered. Then other systems should be considered for the particular application. Finally, cost of the soil nail wall should be considered. Soil nail walls can be used for a variety of soil types and conditions. The most favorable conditions for soil nailing are as follows: The soil should be able to stand unsupported one to two meters high for a minimum of two days when cut vertical or nearly vertical. Also all soil nails within a cross section should be located above the groundwater table. If the soil nails are not located above the groundwater table, the groundwater should not negatively affect the face of the excavation, the bond between the ground and the soil nail itself. Based upon these favorable conditions for soil nailing stiff to hard fine-grained soils which include stiff to hard clays, clayey silts, silty clays, sandy clays, and sandy silts are preferred soils. Sand and gravels which are dense to very dense soils with some apparent cohesion also work well for soil nailing. Weathered rock is also acceptable as long as the rock is weathered evenly throughout (meaning no weakness planes). Finally, glacial soils work well for soil nailing.

A list of unfavorable or difficult soil conditions for soil nailing can include dry, poorly graded cohesion-less soils, soils with a high groundwater table, soils with cobbles and boulders, soft to very soft fine-grained soils, highly corrosive soils, weathered rock with unfavorable weakness planes, and loess. Other difficult conditions include prolonged exposure to freezing temperatures, a climate that has a repeated freeze-and-thaw cycle, and granular soils that are very loose.

Origins

Soil nailing evolved from the New Austrian Tunneling method which is a system for underground excavations in rock. This method consists of passive steel reinforcement in the rock followed by the application of reinforced shotcrete. This concept of combining passive steel reinforcement and shotcrete has also been applied to the stabilization of rock slopes since the early 1960s.

The first application of soil nailing was implemented in 1972 for a railroad widening project near Versailles, France. Soil nails were used to stabilize an 18 m high slope consisting of sandy soil. This method proved to be more cost-effective, while at the same time cut down the construction time when compared to other conventional support methods. Germany was the next country to investigate soil nailing. From 1975 to 1981 the University of Karlsruhe and the construction company Bauer collaborated to establish a research program. This program conducted full scale testing of experimental walls with different configurations and developed analysis procedures for use in design. The United States first used soil nailing in 1976 for the support of a 13.7 m deep foundation excavation in dense silty sands. Soil nailing was implemented in the expansion of The Good Samaritan Hospital in Portland, Oregon. This retaining system was produced in approximately half the time at about 85% of the cost of conventional retaining systems.

Design

After a preliminary analysis of the site, initial designs of the soil nail wall can be begin. This begins with a selection of limit states and design approaches. The two most common limit states used in soil nail wall design is strength limit and service limit states. The strength limit state is the limit state that addresses potential failure mechanisms or collapse states of the soil nail wall system. The service limit state is the limit state that addresses loss of service function resulting from excessive wall deformation and is defined by restrictions in stress, deformation and facing crack width under regular service conditions. The two most common design approaches for soil nail walls are limit state design and service load design.

Initial design considerations include wall layout (wall height and length), soil nail vertical and horizontal spacing, soil nail pattern on wall face, soil nail inclination, soil nail length and distribution, soil nail material and relevant ground properties. With all these variables in the mind of the design engineer the next step is to use simplified charts to preliminarily evaluate nail length and maximum nail force. Nail length, diameter and spacing typically control external and internal stability of the wall. These parameters can be adjusted during design until all external and internal stability requirements are met. After the initial design is completed, final design progresses where the soil nail wall has to be tested for external and internal failure modes, seismic considerations and aesthetic qualities. Drainage, frost penetration and external loads such as wind and hydrostatic forces also have to be determined and included in the final examination of the design. Soil nail walls are not ideal in locations with highly plastic clay soils. Soils with high plasticity, a high liquid limit and low undrained shear strengths are at risk of long-term deformation (creep).

Construction

With the design complete, construction is the next step. Most soil nail wall construction follows a specific procedure. First a cut is excavated and temporary bracing is put in place if necessary. This is done with conventional earth moving equipment and hydraulic drills. Next, holes for the soil nails are drilled at predetermined locations as specified by the design engineer. The equipment used for this step is dependent on the stability of the material in which the soil nail wall is supporting. Rotary or rotary percussive methods using air flush or dry auger methods can be used with stable ground. For unstable ground, single tube and duplex rotary methods with air and water flush or hollow stem auger methods are used. With the holes drilled, the next step is to install and grout the nails into place. After all nails are inserted, a drainage system is put into place. Synthetic drainage mat is placed vertically between the nail heads, which are extended down to the base of the wall where they are most commonly connected to a footing drain. A layer of shotcrete is applied and bearing plates are installed before a final facing is put in place to complete the soil nail wall. Variations of the steps described above may be necessary to accommodate additional preparation tasks or supplementary activities for specific project conditions.

In terms of construction, soil nail walls have a decisive advantage over other alternatives. Soil nail walls require a smaller right-of-way than ground anchor walls and have less environmental impact. Installation of soil nail walls is relatively rapid and typically uses less materials and smaller construction equipment than ground anchor walls.

Cost Comparison

One great advantage of soil nail walls is their cost-effectiveness over other alternatives. When conventional soil nailing construction procedures are used, soil nail walls are much more economical than concrete gravity walls and similarly or more cost effective than ground anchor walls.

Inspection and Performance Monitoring

Inspection activities play a vital role in the production of high-quality soil nail walls because conformance to project plans and specifications should result in a soil nail wall that will perform its intended duty for its designed duration. Inspections usually involve evaluation of the following: conformance of system components to material specification, conformance of construction methods to execution specifications, conformance to short-term performance specifications, and long-term monitoring. Short-term performance specifications are checked with loads tests, which utilize hydraulic jacks and pumps to perform several load applications. Three common load tests for short-term performance are verification or ultimate load tests, proof tests and creep tests. Verification or ultimate load tests are conducted to verify the compliance of the soil nails with pullout capacity and strengths resulting from the contractor's installation method. Proof tests are intended to verify that the contractor's construction procedure has been consistent and that the nails have not been drilled and grouted in a soil zone not tested in the verification stage. Creep tests are performed to ensure that the nail design loads can be safely carried throughout the structure's service life.

Long-term performance monitoring is used to collect data to ensure adequate performance and refine future design practices. Parameters to be measured include vertical and horizontal movement of the wall face, local movements or deterioration of facing elements, drainage to the ground, loads, load distribution and load changes in the nails, temperature and rainfall. These parameters are measured using several specific tools including inclinometers, load cells and strain gauges.

Soil Compaction

In geotechnical engineering, soil compaction is the process in which a stress applied to a soil causes densification as air is displaced from the pores between the soil grains. When stress is applied that causes densification due to water (or other liquid) being displaced from between the soil grains, then consolidation, not compaction, has occurred. Normally, compaction is the result of heavy machinery compressing the soil, but it can also occur due to the passage of (e.g.) animal feet.

In soil science and agronomy, soil compaction is usually a combination of both engineering compaction and consolidation, so may occur due to a lack of water in the soil, the applied stress being internal suction due to water evaporation as well as due to passage of animal feet. Affected soils become less able to absorb rainfall, thus increasing runoff and erosion. Plants have difficulty in compacted soil because the mineral grains are pressed together, leaving little space for air and water, which are essential for root growth. Burrowing animals also find it a hostile environment, because the denser soil is more difficult to penetrate. The ability of a soil to recover from this type of compaction depends on climate, mineralogy and fauna. Soils with high shrink-swell capacity, such as vertisols, recover quickly from compaction where moisture conditions are variable (dry spells shrink the soil, causing it to crack). But clays which do not crack as they dry cannot recover from compaction on their own unless they host ground-dwelling animals such as earthworms — the Cecil soil series is an example.

A crawler-backhoe is here equipped with a narrow sheepsfoot roller to compact the fill over newly placed sewer pipe, forming a stable support for a new road surface.

A compactor/roller fitted with a sheepsfoot drum, operated by U.S. Navy Seabees.

A Hamm vibrating roller with plain drum as used for compacting asphalt and granular soils.

Wacker Neuson vibratory rammer BS 60-2i in action.

In Construction

Soil compaction is a vital part of the construction process. It is used for support of structural entities such as building foundations, roadways, walkways, and earth retaining structures to name a few. For a given soil type certain properties may deem it more or less desirable to perform adequately for a particular circumstance. In general, the preselected soil should have adequate strength, be relatively incompressible so that future settlement is not significant, be stable against volume change as water content or other factors vary, be durable and safe against deterioration, and possess proper permeability.

When an area is to be filled or backfilled the soil is placed in layers called lifts. The ability of the first fill layers to be properly compacted will depend on the condition of the natural material being covered. If unsuitable material is left in place and backfilled, it may compress over a long period under the weight of the earth fill, causing settlement cracks in the fill or in any structure supported by the fill. In order to determine if the natural soil will support the first fill layers, an area can be proofrolled. Proofrolling consists of utilizing a piece heavy construction equipment (typically, heavy compaction equipment or hauling equipment) to roll across the fill site and watching for deflections to be revealed. These areas will be indicated by the development of rutting, pumping, or ground weaving.

To ensure adequate soil compaction is achieved, project specifications will indicate the required soil density or degree of compaction that must be achieved. These specifications are generally recommended by a geotechnical engineer in a geotechnical engineering report.

The soil type - that is, grain-size distributions, shape of the soil grains, specific gravity of soil solids, and amount and type of clay minerals, present - has a great influence on the maximum dry unit weight and optimum moisture content. It also has a great influence on how the materials should be compacted in given situations. Compaction is accomplished by use of heavy equipment. In sands and gravels, the equipment usually vibrates, to cause re-orientation of the soil particles into a denser configuration. In silts and clays, a shccpsfoot roller is frequently used, to create small zones of intense shearing, which drives air out of the soil.

Determination of adequate compaction is done by determining the in-situ density of the soil and comparing it to the maximum density determined by a laboratory test. The most commonly used laboratory test is called the Proctor compaction test and there are two different methods in obtaining the maximum density. They are the standard Proctor and modified Proctor tests; the modified Proctor is more commonly used. For small dams, the standard Proctor may still be the reference.

While soil under structures and pavements needs to be compacted, it is important after construction to decompact areas to be landscaped so that vegetation can grow.

Compaction Methods

There are several means of achieving compaction of a material. Some are more appropriate for soil compaction than others, while some techniques are only suitable for particular soils or soils in particular conditions. Some are more suited to compaction of non-soil materials such as asphalt. Generally, those that can apply significant amounts of shear as well as compressive stress, are most effective.

The available techniques can be classified as:

1. Static - a large stress is slowly applied to the soil and then released.
2. Impact - the stress is applied by dropping a large mass onto the surface of the soil.
3. Vibrating - a stress is applied repeatedly and rapidly via a mechanically driven plate or hammer. Often combined with rolling compaction.
4. Gyrating - a static stress is applied and maintained in one direction while the soil is a subjected to a gyratory motion about the axis of static loading. Limited to laboratory applications.
5. Rolling - a heavy cylinder is rolled over the surface of the soil. Commonly used on sports pitches. Roller-compactors are often fitted with vibratory devices to enhance their effectiveness.
6. Kneading - shear is applied by alternating movement in adjacent positions. An example, combined with rolling compaction, is the 'sheepsfoot' roller used in waste compaction at landfills.

The construction plant available to achieve compaction is extremely varied and is described elsewhere.

Test Methods in Laboratory

Soil compactors are used to perform test methods which cover laboratory compaction methods used to determine the relationship between molding water content and dry unit weight of soils. Soil placed as engineering fill is compacted to a dense state to obtain satisfactory engineering properties such as, shear strength, compressibility, or permeability. In addition, foundation soils are often compacted to improve their engineering properties. Laboratory compaction tests provide the basis for determining the percent compaction and molding water content needed to achieve the required engineering properties, and for controlling construction to assure that the required compaction and water contents are achieved. Test methods such as EN 13286-2, EN 13286-47, ASTM D698, ASTM D1557, AASHTO T99, AASHTO T180, AASHTO T193, BS 1377:4 provide soil compaction testing procedures.

Soil Bioengineering

Soil bioengineering is a discipline of civil engineering. It pursues technological, ecological, economic as well as design goals and seeks to achieve these primarily by making use of living materials, i.e. seeds, plants, part of plants and plant communities, and employing them in near–natural constructions while exploiting the manifold abilities inherent in plants. Soil bioengineering may sometimes be a substitute for classical engineering works; however, in most cases it is a meaningful and necessary method of complementing the latter. Its application suggests itself in all fields of soil and hydraulic engineering, especially for slope and embankment stabilization and erosion control.

Soil bioengineering is the use of living plant materials to provide some engineering function. Soil bioengineering is an effective tool for treatment of a variety of unstable and / or eroding sites. Soil bioengineering techniques have been used for many centuries. More recently Schiechtl (1980) has encouraged the use of soil bioengineering with a variety of European examples. Soil bioengineering is now widely practiced throughout the world for the treatment of erosion and unstable slopes.

Fields of Application and Plants for Soil Bioengineering Control Works

Soil Bioengineering methods can be applied wherever the plants which are used as living building materials are able to grow well and develop. This is the case in tropical, subtropical and temperate zones whereas there are obvious limits in dry and cold regions, i.e. where arid, semi–arid and frost zones prevail. In exceptional cases, lack of water may be compensated for by watering or irrigation. In Europe, dry conditions limiting application exist in the Mediterranean as well as in some inner alpine and eastern European snowy regions. However, limits are most frequently imposed in alpine and arctic regions. These can usually be clearly noticed by the limited growth of woody plants (forest, tree and shrub lines) and the upper limits of closed turf cover. The more impoverished a region is in species, the less suited it is for the application of bioengineering methods.

Functions and Effects of Soil Bioengineering Structures

Technical Functions

- protection of soil surface from erosion by wind, precipitation, frost and flowing water
- protection from rock fall
- elimination or binding of destructive mechanical forces
- reduction of flow velocity along banks
- surface and/or deep soil cohesion and stabilization
- drainage

- protection from wind
- aiding the deposition of snow, drift sand and sediments
- increasing soil roughness and thus preventing avalanche release

Apart from these, ecological functions are gaining in importance, particularly as these can be fulfilled to a very limited extent only by classical engineering constructions.

Ecological Functions

- improvement of water regime by improved soil interception and storage capability as well as water
- consumption by plants
- soil drainage
- protection from wind
- protection from ambient air pollution
- mechanical soil amelioration by the roots of plants
- balancing of temperature conditions in near–ground layers of air and in the soil
- shading
- improvement of nutrient content in the soil and thus of soil fertility on previously raw soils
- balancing of snow deposits
- noise protection
- yield increase on neighbouring cropland

Landscaping Functions

- healing of wounds inflicted on the landscape by disasters and humans (exploitation of mineral resources, construction work, deposition of overburden, tunnel excavation material, industrial and domestic waste)
- integration of structures into the landscape
- concealment of offending structures
- enrichment of the landscape by creating new features and structures, shapes and colours of vegetation

Economic Effects

Bioengineering control works are not always necessarily cheaper in construction when compared to classical engineering structures. However, when taking into account their lifetime including their service and maintenance, they will normally turn out to be more economical. Their special advantages are:

- lower construction costs compared to "hard" constructions
- lower maintenance and rehabilitation costs
- creation of useful green areas and woody plant populations on previously derelict land
- Useful for income generation

The result of soil bioengineering protection works are living systems which develop further and maintain their balance by natural succession (i.e. by dynamic self–control, without artificial input of energy). If the right living but also non–living building materials and the appropriate types of construction are chosen, exceptionally high sustainability requiring little maintenance effort can be achieved.

Pedometrics

Pedometrics is the application of mathematical and statistical methods for the study of the distribution and genesis of soils.

Pedometrics is a portmanteau of the Greek roots *pedos* (soil) and *metron* (measurement). Measurement in this case is restricted to mathematical and statistical methods as it relates to pedology, the branch of soil science that studies soil in its natural setting.

Pedometrics addresses soil related problems when there is uncertainty due to deterministic or stochastic variation, vagueness and lack of knowledge of soil properties and processes. It relies upon mathematical, statistical, and numerical methods and includes numerical approaches to classification to deal with a supposed deterministic variation. Simulation models incorporate uncertainty by adopting chaos theory, statistical distribution, or fuzzy logic.

Pedometrics addresses pedology from the perspective of emerging scientific fields such as wavelets analysis, fuzzy set theory and data mining in soil data modelling applications. The advance of pedometrics is also linked to improvements in remote and close-range sensing.

Pedometrics society logo.

References

- McCarthy, David F. (2007). Essentials of Soil Mechanics and Foundations. Upper Saddle River, NJ: Pearson Prentice Hall. p. 595. ISBN 0-13-114560-6.
- McCarthy, David F. (2007). Essentials of Soil Mechanics and Foundations. Upper Saddle River, NJ: Pearson Prentice Hall. pp. 601–602. ISBN 0-13-114560-6.
- McCarthy, David F. (2007). Essentials of Soil Mechanics and Foundations. Upper Saddle River, NJ: Pearson Prentice Hall. p. 602. ISBN 0-13-114560-6.
- Blonquist, J. M. (2005). "A time domain transmission sensor with TDR performance characteristics" (PDF). Journal of Hydrology. 314: 235–245. doi:10.1016/j.jhydrol.2005.04.005. Retrieved 31 Jan 2016.

Permissions

All chapters in this book are published with permission under the Creative Commons Attribution Share Alike License or equivalent. Every chapter published in this book has been scrutinized by our experts. Their significance has been extensively debated. The topics covered herein carry significant information for a comprehensive understanding. They may even be implemented as practical applications or may be referred to as a beginning point for further studies.

We would like to thank the editorial team for lending their expertise to make the book truly unique. They have played a crucial role in the development of this book. Without their invaluable contributions this book wouldn't have been possible. They have made vital efforts to compile up to date information on the varied aspects of this subject to make this book a valuable addition to the collection of many professionals and students.

This book was conceptualized with the vision of imparting up-to-date and integrated information in this field. To ensure the same, a matchless editorial board was set up. Every individual on the board went through rigorous rounds of assessment to prove their worth. After which they invested a large part of their time researching and compiling the most relevant data for our readers.

The editorial board has been involved in producing this book since its inception. They have spent rigorous hours researching and exploring the diverse topics which have resulted in the successful publishing of this book. They have passed on their knowledge of decades through this book. To expedite this challenging task, the publisher supported the team at every step. A small team of assistant editors was also appointed to further simplify the editing procedure and attain best results for the readers.

Apart from the editorial board, the designing team has also invested a significant amount of their time in understanding the subject and creating the most relevant covers. They scrutinized every image to scout for the most suitable representation of the subject and create an appropriate cover for the book.

The publishing team has been an ardent support to the editorial, designing and production team. Their endless efforts to recruit the best for this project, has resulted in the accomplishment of this book. They are a veteran in the field of academics and their pool of knowledge is as vast as their experience in printing. Their expertise and guidance has proved useful at every step. Their uncompromising quality standards have made this book an exceptional effort. Their encouragement from time to time has been an inspiration for everyone.

The publisher and the editorial board hope that this book will prove to be a valuable piece of knowledge for students, practitioners and scholars across the globe.

Index

S

T

W